SpringerBriefs in Space Development

Series Editor

Joseph N. Pelton Jr., Arlington, USA

More information about this series at http://www.springer.com/series/10058

Erik Seedhouse

Space Radiation and Astronaut Safety

INTERNATIONAL®
SPACE UNIVERSITY

Springer

Erik Seedhouse
Applied Aviation Sciences
Embry-Riddle Aeronautical University
Daytona Beach, FL, USA

ISSN 2191-8171 ISSN 2191-818X (electronic)
SpringerBriefs in Space Development
ISBN 978-3-319-74614-2 ISBN 978-3-319-74615-9 (eBook)
https://doi.org/10.1007/978-3-319-74615-9

Library of Congress Control Number: 2018933376

Printed on acid-free paper

This Springer imprint is published by the registered company Springer International Publishing AG part of Springer Nature.
The registered company address is: Gewerbestrasse 11, 6330 Cham, Switzerland

This Springer book is published in collaboration with the International Space University. At its central campus in Strasbourg, France, and at various locations around the world, the ISU provides graduate-level training to the future leaders of the global space community. The university offers a two-month Space Studies Program, a five-week Southern Hemisphere Program, a one-year Executive MBA and a one-year Master's program related to space science, space engineering, systems engineering, space policy and law, business and management, and space and society.

These programs give international graduate students and young space professionals the opportunity to learn while solving complex problems in an intercultural environment. Since its founding in 1987, the International Space University has graduated more than 3000 students from 100 countries, creating an international network of professionals and leaders. ISU faculty and lecturers from around the world have published hundreds of books and articles on space exploration, applications, science, and development.

Contents

About the Author

Erik Seedhouse is a highly published author. After completing his first degree, he joined the Second Battalion of the Parachute Regiment. During his time in the "Para's," Erik spent six months in Belize, where he was trained in the art of jungle warfare. Later, he spent several months learning the intricacies of desert warfare in Cyprus. He made 30+ jumps from a C130 aircraft, performed more than 200 helicopter abseils, and fired more light anti-tank weapons than he cares to remember!

Upon returning to academia, the author embarked upon a master's degree which he supported by winning prize money in 100-km running races. After placing third in the World 100-km Championships in 1992, Erik turned to ultra-distance triathlon, winning the World Endurance Triathlon Championships in 1995 and 1996. For good measure he won the World Double Ironman Championships in 1995 and the infamous Decatriathlon, an event requiring competitors to swim 38 km, cycle 1800 km, and run 422 km. Non-stop!

In 1996, Erik pursued his PhD at the German Space Agency's Institute for Space Medicine. While studying he found time to win Ultraman Hawai'i and the European Ultraman Championships as well as completing Race Across America. Due to his success as the world's leading ultra-distance triathlete Erik was featured in dozens of magazine and television interviews. In 1997 *GQ* magazine named him the "Fittest Man in the World."

Erik's PhD in space medicine and background in space life sciences provided him with a keen insight into and understanding of the medical problems faced by long duration astronauts. In 1999 Erik took a research job at Simon Fraser University. In 2005 the author worked as an astronaut training consultant for Bigelow Aerospace. Between 2008 and 2013 he served as director of Canada's manned centrifuge and hypobaric operations. In 2009 he was one of the final 30 candidates in the Canadian Space Agency's Astronaut Recruitment Campaign. Erik has a dream job as an assistant professor at Embry-Riddle Aeronautical University in Daytona Beach, Florida. In his spare time he works as an astronaut instructor for Project PoSSUM, occasional film consultant to Hollywood, a professional speaker, triathlon coach, and author. This is his 27th book. When not enjoying the sun and rocket launches on Florida's Space Coast with his fiancée, Alice, he divides his time between his second home in Sandefjord and Mauna Lani on the Big Island of Hawai'i.

Abbreviations

AHARS	As high as reasonably acceptable
ALARA	As low as reasonably achievable
ARS	Acute radiation syndrome
AST	Attentional set-shifting task
BEO	Beyond earth orbit
BFO	Blood forming organ
BMD	Bone mineral density
CME	Coronal mass ejection
CMO	Crew medical officer
CNS	Central nervous system
CPDS	Charged particle detector
ESA	European Space Agency
EVA	Extravehicular activity
EV-CPDS	Extravehicular charged particle directional spectrometer
GCR	Galactic cosmic radiation
GOES	Geostationary operational environmental satellite
Gy	Gray
HSC	Hematopoietic stem cells
ICRP	International Council on Radiation Protection
ISS	International Space Station
ITS	Interplanetary transport system
IV-CPDS	Intravehicular charged particle directional spectrometer
JSC	Johnson Space Center
LEO	Low earth orbit
LET	Linear energy transfer
LOC	Loss of crew
LOM	Loss of mission
LPA	Lysophosphatidic acid
LSAH	Longitudinal survey of astronaut health
MORD	Medical operations requirements document
MPCV	Multi-purpose crew vehicle

MSL	Mars Science Laboratory
NAD	Nicotinamide adenine dinucleotide
NAS	National Academy of Sciences
NCR	National Cancer Institute
NCRP	National Council on Radiation Protection
NHP	Non-human primate
NMN	Nicotinamide mononucleotide
NOAA	National Oceanic Atmospheric Administration
PEL	Permissible exposure limits
PSD	Positron sensitive detector
RAD	Radiation assessment detector
RAM	Radiation area monitor
RBE	Relative biological effectiveness
REID	Risk of exposure-induced death
SAA	South Atlantic anomaly
SEC	Space Environment Center
SPE	Solar particle event
SRAG	Space radiation assessment group
SRHO	Space radiation health officer
TBI	Total body irradiation
TEPC	Tissue equivalent proportional counter
TLD	Thermoluminescent detector
UNSCEAR	United Nations Scientific Committee on the Effects of Atomic Radiation

Chapter 1
Radiation: A Primer

One of the most challenging parts for the human journey to Mars is the risk of radiation exposure and the inflight and long-term health consequences of the exposure. This ionizing radiation travels through living tissues, depositing energy that causes structural damage to DNA and alters many cellular processes.

—NASA Space Radiation Element Scientist Lisa Simonsen, Ph.D.

Galactic cosmic ray exposure can devastate a cell's nucleus and cause mutations that can result in cancers. We learned the damaged cells send signals to the surrounding, unaffected cells and likely modify the tissues' microenvironments. Those signals seem to inspire the healthy cells to mutate, thereby causing additional tumors or cancers.

—Dr. Francis Cucinotta, University of Nevada, Las Vegas

Exposure to these particles can lead to a range of potential central nervous system complications that can occur during and persist long after actual space travel—such as various performance decrements, memory deficits, anxiety, depression, and impaired decision-making. Many of these adverse consequences to cognition may continue and progress throughout life.

—Dr. Charles Limoli, radiation oncology professor at the University of California, Irvine

As long as there have been astronauts there has been talk of a manned mission to Mars. Hardly a week goes by without an announcement of another humans-to-Mars initiative. Over the years the public has been introduced to Inspiration Mars, Mars One, and SpaceX's Interplanetary Transport System. Most such announcements garner plenty of press before dying a slow and natural death, but each mission shares one common denominator: they are either oblivious to (such as in the case of Mars One) or choose to ignore (Mars One again) the dangers of space radiation (Figs. 1.1 and 1.2). So let's take a look at what radiation sources are out there. We'll begin with galactic cosmic rays (GCR)

E. Seedhouse, *Space Radiation and Astronaut Safety*, SpringerBriefs in Space Development, https://doi.org/10.1007/978-3-319-74615-9_1

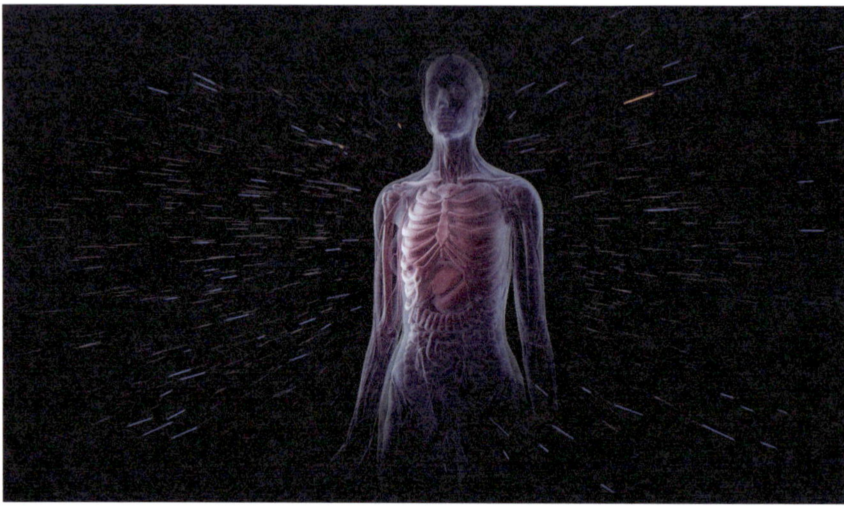

Fig. 1.1 Radiation damages every system in the body. (Image courtesy of NASA)

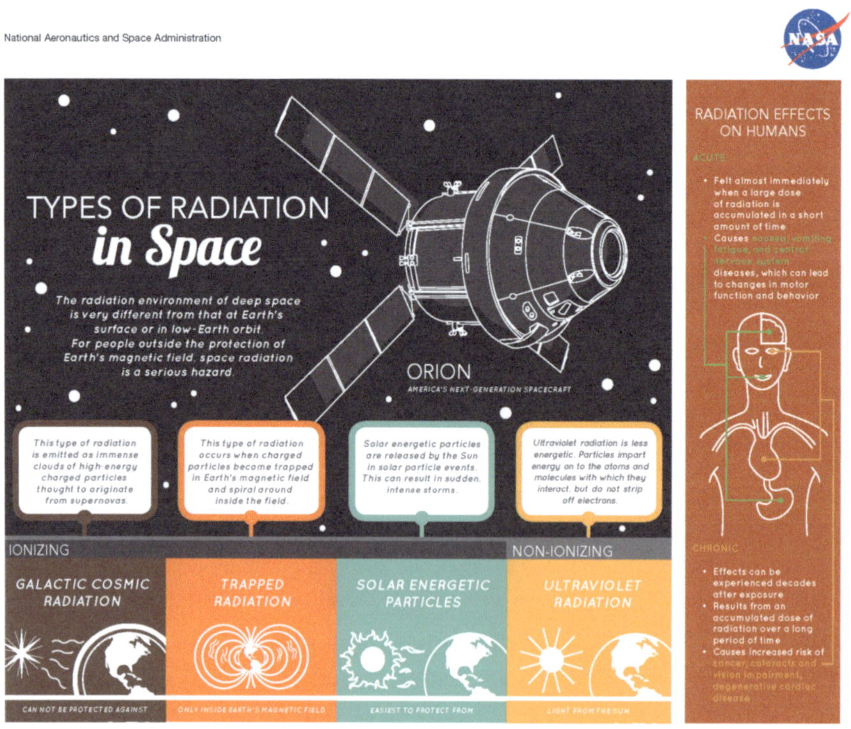

Fig. 1.2 Types of space radiation. (Image courtesy of NASA)

Galactic Cosmic Rays

Why all the fuss over this particular type of radiation? GCRs (Fig. 1.3) are unique for the simple reason it is nigh on impossible to shield against them, and here's why. GCRs, which are highly energetic charged particles that originate outside the Solar System, bullet along at close to the speed of light, propelled through space by the force of exploding stars [1, 2]. This means these particles have tremendous energy. The mass of some of these particles, such as iron, combined with their phenomenal speed, enable them to slice through the walls of a spacecraft like the proverbial hot knife through butter. And astronauts exposed to too much cosmic radiation will be at high risk of developing cancer. And not just any cancer. Here's what Dr. Francis Cucinotta of the University of Nevada in Las Vegas [3], one of the world's leading experts on the physiological effects of space radiation, said: "The type of tumors that cosmic ray ions make are more aggressive than what we get from other radiation."

GCR are comprised almost entirely of protons and alpha particles (about 99%), with the remaining 1% composed of heavy nuclei such as lithium, beryllium and boron. GCR composed of charged nuclei that are heavier than helium are termed HZE ions. HZE ions are scarce, but because these particles are so highly charged and so heavy, they contribute to a large proportion of an astronaut's radiation dose.

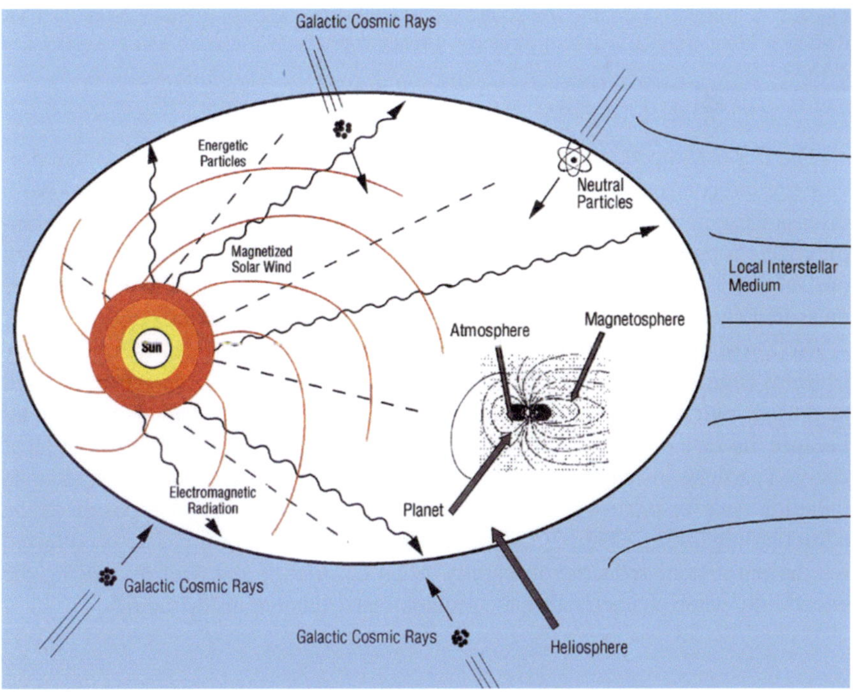

Fig. 1.3 Galactic cosmic radiation. (Image courtesy of NASA)

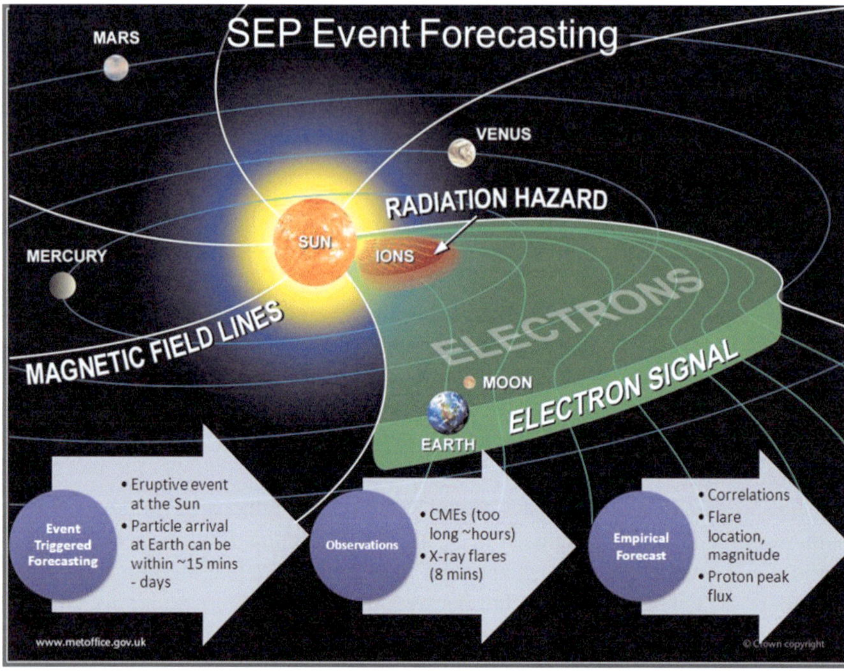

Fig. 1.4 Astronauts conducting spacewalks are monitored for radiation exposure using the EVA Radiation Monitoring (EVARM) instrument. (Image courtesy of Canadian Space Agency and NASA)

The ISS is equipped with a suite of active radiation monitors (Fig. 1.4) that provide ground controllers and crew with cumulative exposures and dose rates. We'll discuss these in more detail in Chap. 5, so what follows is an overview. One example of the monitors carried on ISS is a spectrometer that provides time-resolved measurements of GCR and the components of GCR such as neutrons, which are measured via neutron spectroscopy. Why are neutrons so important?

Well, results from various studies over the years have revealed that secondary neutrons contribute up to 30% of the total radiation dose that astronauts are exposed to during their tour of duty [4]. Charged particle monitoring is also important because this data provides information about the direction-dependent distribution of charged particles inside the ISS [5]. This data can then be used to calculate accurate radiation data for organ exposure and the consequent risk of that exposure to the particular organ (see Chap. 2). All in all, the accumulation of this data is collected for the purpose of reducing ambiguity when calculating risk and decreasing the uncertainty when characterizing the radiation environment inside the ISS.

Solar Particle Events

The next component of space radiation are solar particle events (SPEs, Fig. 1.5). SPEs are comprised of energetic electrons, protons, and alpha particles that are accelerated to speeds approaching light speed by shockwaves that precede coronal mass ejections (CMEs), which occur near solar flare sites. The most energetic SPEs may arrive in low Earth orbit (LEO) within 20 or 30 min of the event and may cause significant effects in Earth's atmosphere. Fortunately, those effects are predictable—to a degree. That's because the Sun's activity follows an 11-year cycle that consists of four inactive years (solar minimum) and seven active years (solar maximum).

During solar maximum, the Sun generates about three CMEs per day, compared to one every five days during solar minimum. These events and the timing of them are significant because a typical CME contains billions of tons of matter, and the shockwaves associated with CMEs may generate magnetic storms that can impact those residing in LEO. During a large SPE, the fluence of protons may exceed tens of millions of electron volts (MeV) and may significantly increase the radiation dose for crews in LEO. Needless to say, such energies impose significant operational constraints upon mission planners. Unfortunately there are no ways to reliably predict when an SPE may strike; the best scientists can do is to study (Fig. 1.6) the relationship between SPE intensity and the parameters of shock and plasma, to better understand the conditions when these events occur [6].

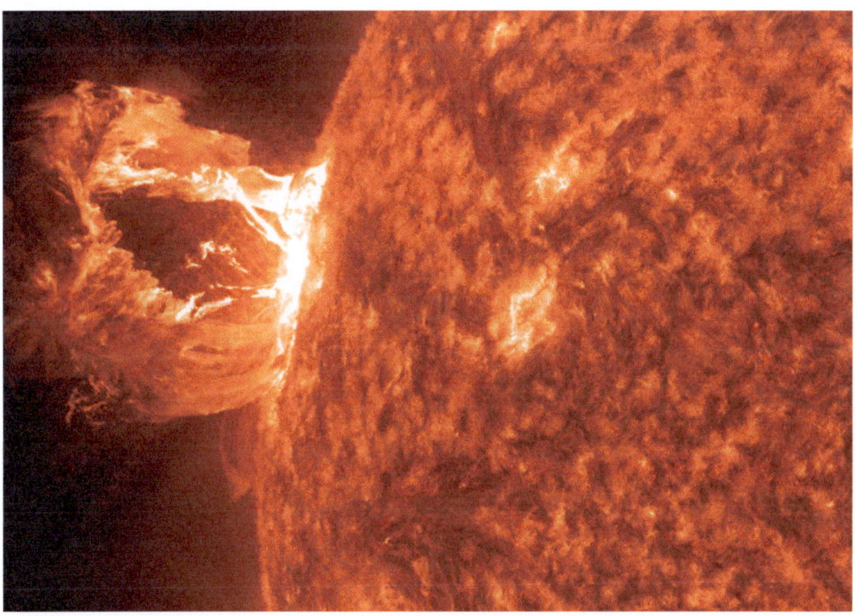

Fig. 1.5 Solar particle events eject billions of tons of material into space. (Image courtesy of NASA)

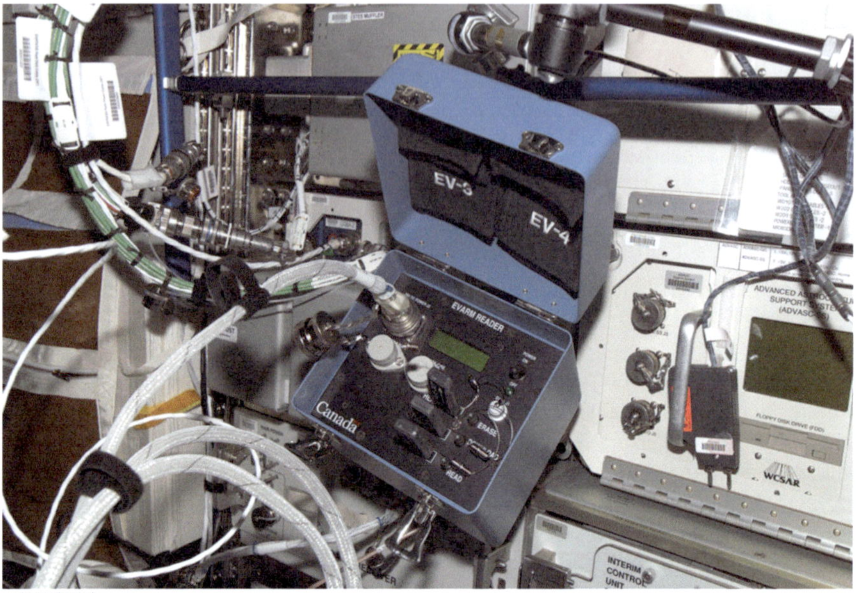

ISS006E48915

Fig. 1.6 Scientists can study SPE's (also known as SEPs) in an attempt to predict solar weather. (Image courtesy of NASA)

Thanks to measurements by the Geostationary Operational Environmental Satellite (GOES) system, scientists have been able to classify the size of SPEs. Most SPEs are classified as A, B, C, M or X, with each class having a peak flux ten times greater than the previous one. A linear scale exists within each class, which means that a flare rated X2 is twice as powerful as a flare rated X1 (or four times as powerful as one rated M5). The flares that generally are the most disruptive are those in the M and X categories. To estimate radiation exposure, scientists use the spectra of SPEs to estimate particle intensities. This information is then used to model radiation exposures, but due to the unpredictable characteristics of particle acceleration and propagation of these particles in interplanetary space [7], the spectra of many SPEs are difficult to determine. To partially overcome this problem, scientists use algorithms that sequentially interpolate measured data points. This information may also be used to model radiation exposure. To determine acute radiation risks caused by SPE exposure, a projection code was developed by NASA [8]. Known as the Acute Radiation Risk/BRYNTRRN Organ Dose (ARRBOD) projection code, this software can be used to analyze SPEs. This information, combined with information obtained from human phantom models and dosimetry (Chap. 5), can be used to estimate resultant organ doses in those astronauts who encounter an SPE event. This information in turn can be used to model the severity of acute radiation sickness (Chap. 7) and the consequent effects that may include vomiting, nausea, and weakness.

Fig. 1.7 Mars science laboratory. (Image courtesy of NASA)

In practice, the interplanetary SPE environment has been characterized by monitors such as the Radiation Assessment Detector (RAD), which was carried on board the Mars Science Laboratory (MSL, Fig. 1.7). The RAD recorded three SPEs (January 23, 2012, March 7, 2012, and May 17, 2012) [9, 10]. None of these events were major SPEs, but timing is everything in spaceflight; between *Apollo 16* and *17,* one of the most powerful SPE's occurred. Had there been astronauts en-route to the Moon at the time, crewmembers would have absorbed lethal radiation doses within 10 h of launch.

Radiation Doses on Board the International Space Station

The ISS (Fig. 1.8) orbits at an altitude of about 400 km (at perigee it is at 400 km and at apogee it orbits at 408 km), which is within the Earth's magnetosphere, but this still means astronauts are exposed to high fluxes of ionizing radiation, the primary sources of which are GCRs. In addition to GCRs, ISS crews are exposed to particles trapped in the Van Allen Belts and SPEs [2, 11, 12]. Another key element in calculating the radiation exposure of astronauts on board the orbiting outpost is the orbital inclination of the ISS, which is 52°. This inclination means that the station passes through the South Atlantic Anomaly (SAA, Fig. 1.9) every day.

The SAA, which is located east of Argentina, is a region characterized by an anomaly in Earth's geomagnetic field that results in energetic particles penetrating to lower altitudes than normal. This means that when the ISS passes through the SAA astronauts are exposed to higher levels of ionizing radiation. Astronauts are already at risk of developing cancer simply by traveling to space. Astronauts on the

Fig. 1.8 International space station. (Image courtesy of NASA)

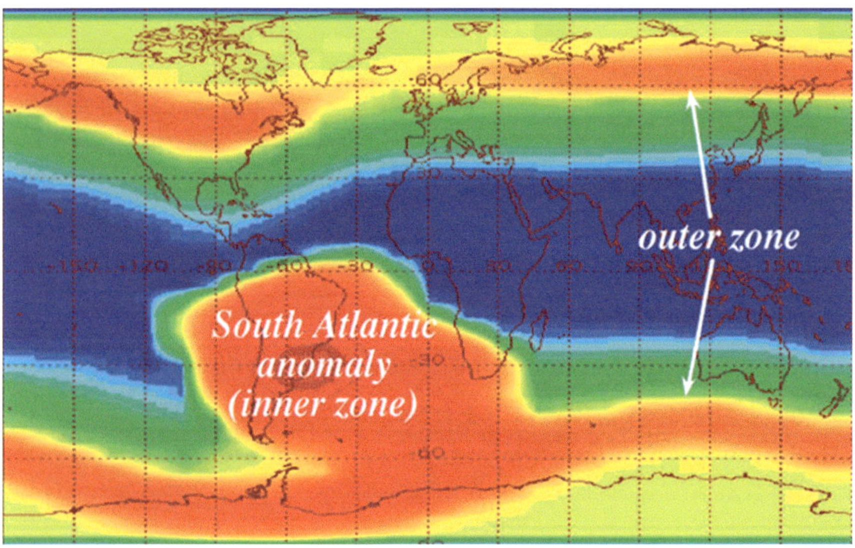

Fig. 1.9 South Atlantic anomaly. (Image courtesy of NASA and NOAA)

ISS receive 80 milliSieverts (mSv) during a 6-month period, whereas people on Earth receive 2 mSv a year. Another comparative metric is the chest radiograph dose, which is 0.02 μSv/h. And if you happen to spend a lot of time flying commercial the radiation dose per hour is about 0.3–5.7 μSv. Spend too much time in orbit, and astronauts hit their career radiation limit, which equates to an increase of 3% risk (sidebar) of developing cancer in their lifetime. Any astronaut heading for Mars will be exposed to between 1 and 5 Sv and therefore be guaranteed to hit—and

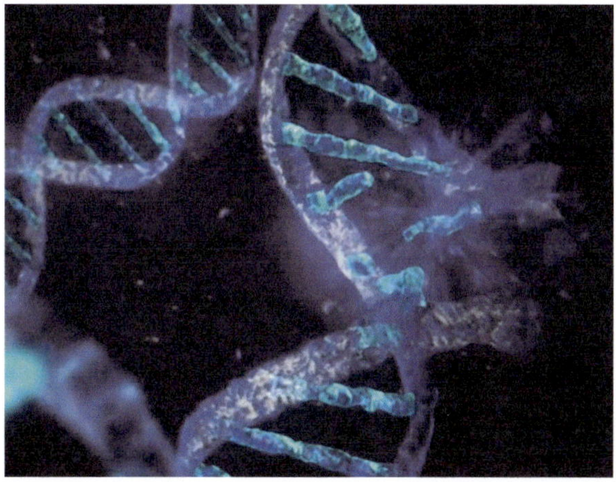

Fig. 1.10 Space radiation can cause serious and lasting damage to genetic material. (Image courtesy of NASA)

exceed—that limit by virtue of being exposed to killer radiation for the best part of 2 years. But remember, this is not just any type of radiation. This radiation isn't stopped by shielding, and there is a lot of it. Out there in deep space, it would only take about three days for every cell in every crewmember to be hit by a high-energy proton. Now, for some cells, it isn't much of a problem, but when these careening nuclei hit important cells such as DNA, mutations may result (Fig. 1.10) [13, 14].

Radiation Risk

In the United States, the incidence rate of cancer is 38.5% (according to the National Cancer Institute based on statistics between 2008 and 2012). If you exposed 100 people (which is the capacity of SpaceX's Mars-bound Interplanetary Transport System incidentally) to the amount of radiation that Mars astronauts will be exposed to, 61 of them would be diagnosed with cancer. By virtue of the unique characteristics of GCR, these cancers would typically be lung, breast and colorectal cancers, meaning half of these astronauts would die. Scientists have modeled the dangers of GCRs during a manned Mars mission and have calculated that exposure to radiation on such a trip would shorten an astronaut's lifespan by between 15 and 24 years.

Improving Crew Health

It sounds like a lot of doom and gloom, which is why experts such as Dr. Cucinotta [3] recommend that space agencies gather a lot more data about how GCR affects crew health and how these effects may be mitigated. For example, some of the immune responses caused by GCR exposure are similar to those in inflammatory diseases. In such cases, oxidants are produced that change intercellular signaling, but it is possible that non-steroidal anti-inflammatory drugs could be taken by astronauts during their Mars mission to mitigate some of the effects of cancer [15]. Another more serious effect of GCR is that it may accelerate the onset of symptoms similar to those exhibited by Alzheimer's patients [16]:

> *Galactic cosmic radiation poses a significant threat to future astronauts. The possibility that radiation exposure in space may give rise to health problems such as cancer has long been recognized. However, this study shows for the first time that exposure to radiation levels equivalent to a mission to Mars could produce cognitive problems and speed up changes in the brain that are associated with Alzheimer's disease.*
> —Professor M. Kerry O'Banion, M.D., Ph.D., University of Rochester Medical Center Department of Neurobiology and Anatomy

O'Banion's research focused on how GCR affects the central nervous system (CNS), and the news was less than rosy. Much of his research has been conducted at the NASA Space Radiation Laboratory at Brookhaven National Laboratory on Long Island. It's a place that was chosen for its accelerators, which can collide matter at extremely high speeds, thereby replicating what happens in space. O'Banion's research examined the effect of GCR-equivalent radiation on cognitive function. To do this, mice were exposed to various doses of radiation comparable to levels that Mars-bound astronauts would be subjected to. The mice were subject to recall tests, and researchers found that mice exposed to radiation were more likely to fail the tests, a finding that indicated cognitive impairment. Dr. O'Banion says: "Because iron particles pack a bigger wallop it is extremely difficult from an engineering perspective to effectively shield against them. One would have to essentially wrap a spacecraft in a 6-foot block of lead or concrete."

After examining the mice more closely, researchers found that the brains of the mice exhibited signs of vascular alteration and larger than normal amount of beta amyloid, which happens to be a signature of Alzheimer's. The researchers concluded that exposing astronauts to GCR for extended duration might accelerate Alzheimer's. So what can astronauts do? Well, they can shield themselves from the radiation, a subject discussed in Chap. 6, and they can monitor their exposure to radiation while they are inside the vehicle.

Intravehicular Radiation

On board the ISS, passive dosimeters (Fig. 1.11) are located in each pressurized module [4]. These dosimeters measure time-integrated absorbed doses that change according to the station's altitude and position in the solar cycle. The requirements of these dosimeters are defined in the International Space Station Medical Operations Requirements Document (ISS MORD SSP 50260), a document that states the dose limits and radiation exposure practices across all mission phases. These radiation monitors must perform the following functions:

1. Measure cumulative radiation dose.
2. Downlink linear energy transfer data
3. Provide direction-dependent distribution radiation data for inside and outside the ISS.

Fig. 1.11 A radiation area monitor deployed in the ISS. (Image courtesy of NASA)

4. Downlink data for frequent analysis.
5. Downlink dose-rate from charged particle monitoring equipment
6. Alert the crew when exposure rates exceed set threshold

We'll talk more about these monitors later in this brief but for now that concludes our primer on radiation. Before delving deeper on the subject of radiation and astronaut safety it is helpful to provide a structure and outline to this brief.

Purpose of This Brief

The purpose of this brief is to provide a general overview of the hazards of radiation to crews working in space. The corporation SpaceX (Fig. 1.12) is planning on sending a manned mission to Mars before the end of the 2020s, and the company may be followed by NASA shortly thereafter. As preparations for these missions continue, the European Space Agency (ESA) is developing a mission architecture for establishing a permanent presence on the Moon (Fig. 1.13). An important consideration in each of these ventures is protecting crews against exposure to radiation. This is because exposure to radiation may result in long-term health consequences for astronauts on such missions. To help with these concerns, NASA is engaged in research aimed at better understanding the risks of space radiation, evaluating radiation shielding requirements, and recommending a strategic plan for developing appropriate mitigation capabilities. This book presents an assessment of the current

Fig. 1.12 Artist's impression of SpaceX's Dragon V2 arriving at Mars. (Image courtesy of SpaceX)

Fig. 1.13 The European Space Agency has proposed a lunar village to be established by the late 2020s. (Image courtesy of ESA)

knowledge of the radiation environment. It also examines the effects of radiation on biological systems and mission equipment and provides an analysis of current plans for radiation protection.

Structure of This Brief

This introductory chapter has provided an overview of the types of radiation and some of the ways radiation is measured and monitored on board the ISS. Chapter 2 delves into the biomedical short- and long-term consequences of being exposed to space radiation with particular emphasis on mechanisms used by the body to repair this damage. Chapter 3 examines the various ways limits are set by the space agencies and what risk levels are deemed acceptable. These limits are set against terrestrial exposure levels, which are 80 times lower than the level astronauts are exposed to in orbit. These exposure levels are referred to throughout the book to frame the extreme risk that astronauts venturing beyond Earth orbit will need to accept. Many proponents of a manned Mars mission support their argument by pointing out that since astronauts on board ISS spend six months in orbit a six month trip to Mars shouldn't pose any problems. The reality is different. Whereas astronauts on board the ISS are exposed to 80 mSv during a 6-month increment, a Mars-bound astronaut will be exposed to 3.5 Sv. Not only that, but astronauts bound for the Red Planet will be exposed to the same amount of radiation on the return trip, and we haven't even discussed the exposure accrued during the surface stay. In short, the theme of this book is that radiation is perhaps the greatest challenge that must be overcome before sending crews for extended stays in deep space. It is why NASA has an entire group dedicated to assessing the effects of radiation on its crews. This group—the Space Radiation Assessment Group—is introduced in Chap. 4.

> *The primary means by which radiation effects cells is by damaging DNA—breaks in strands could be experienced. DNA bases (adenine, guanine, cytosine, and thymine) can also be knocked out. The cell will make an attempt to repair these damages. Sometimes it's effective and sometimes it's not, and sometimes it can be mis-repaired. Genes that have been mis-repaired can become mutations, and the accumulation of these mutations over time can potentially lead to cancer.*
>
> —Dr. Peter Guida, liaison biologist for NASA Space Radiation Laboratory

Chapter 5 describes some of the ways radiation is measured inside and outside the ISS. The reader is introduced to the various dosimeters space agencies use to track radiation, analyze it, and make recommendations on crew health based on dosimetric data. Chapter 6 delves into the various ways engineers can protect astronauts from radiation. Accepted shielding materials such as polyethylene and water are discussed, and the reader is introduced to more exotic shielding options such as magnetic and electrical shielding. Chapter 7 deals with the challenges of dealing with acute radiation sickness (ARS), the syndromes that comprise ARS, while Chap. 8 discusses how astronauts might deal with the consequences of being exposed to high radiation levels. Pharmacological countermeasures are dealt with in Chap. 9, which discusses candidate radioprotectors and examines the process of radiation injury and repair. Finally, in Chap. 10, the reader is introduced to methods that NASA might implement to ensure astronauts embarked on Mars missions are as radiation-resistant as possible.

References

1. Akopova AB, Manaseryan MM, Melkonyan AA, Tatikyan SS, Potapov Y. Radiation measurement on the International Space Station. Radiat. Meas. 2005;39:225–8.
2. Kodaira S, Kawashima H, Kitamura H, Kurano M, Uchihori Y, Yasuda N, Ogura K, Kobayashi I, Suzuki A, Koguchi Y, Akatov YA, Shurshakov VA, Tolochek RV, Krasheninnikova TK,

Ukraintsev AD, Gureeva EA, Kuznetsov VN, Benton ER. Analysis of radiation dose variations measured by passive dosimeters onboard the International Space Station during the solar quiet period (2007-2008). Radiat. Meas. 2013;49:95–102.
3. Cucinotta FA, Kim MY, Chappell L. Space radiation cancer risk projections and uncertainties-2012. NASA TP 2013-217375, 2013.
4. Getselev I, Rumin S, Sobolevsky N, Ufimtsev M, Podzolko M. Absorbed dose of secondary neutrons from galactic cosmic rays inside the international space station. Adv. Space Res. 2004;34:1429–32.
5. Reitz G, Berger T, Bilski P, Facius R, Hajek M, Petrov VP, Puchalska M, Zhou D, Bossler J, Akatov YA, Shurshakov VA, Olko P, Ptaszliewicz M, Bergmann R, Fugger M, Vana N, Beaujean R, Burmeister S, Bartlett D, Hager L, Palfalvi JK, Szabó J, O'Sullivan D, Kitamura H, Uchihori Y, Yasuda N, Nagamatsu A, Tawara H, Benton ER, Gaza R, McKeever SW, Sawakuchi G, Yukihara EG, Cucinotta FA, Semones E, Zapp EN, Miller J, Dettmann J. Astronaut's organ doses inferred from measurements in a human phantom outside the International Space Station. Radiat. Res. 2009;171(2):225–35.
6. Reames DV. The two sources of solar energetic particles. Space Sci. Rev. 2013;175:53–92.
7. Miroshnichenko AE, et al. Nonradiating anapole modes in dielectric nanoparticles. Nat. Commun. 2015;6:8069.
8. Kim M-H, Hu S, Nounu HN, Cucinotta FA. Development of graphical user interface for ARRBOD (acute radiation risk and BRYNTRN organ dose projection). Hanover: Center for Aerospace Information; 2010. NASA TP-2010-216116.
9. Zeitlin C, Hassler DM, Cucinotta FA, Ehresmann B, Wimmer-Schweingruber RF, Brinza DE, et al. Measurements of energetic particle radiation in transit to Mars on the Mars science laboratory. Science. 2013;340:1080–4.
10. Hassler DM, Zeitlin C, Wimmer-Schweingruber RF, Ehresmann B, Rafkin S, Eigenbrode JL, et al. Mars' surface radiation environment measured with the Mars Science Laboratory's Curiosity Rover. Science. 2014;343:1244797.
11. Dietze G, et al. Assessment of astronaut exposures in space. ICRP publication 123. Ann. ICRP. 2013;42:2013.
12. Cucinotta FA, Kim MY, Willingham V, George KA. Physical and biological organ dosimetry analysis for international space station astronauts. Radiat. Res. 2008;170(1):127–38.
13. Dicello J, Christian A, Cucinotta F, Gridley D, Kathirithamby R, Mann J, et al. In vivo mammary tumourigenesis in the Sprague-Dawley rat and microdosimetric correlates. Phys. Med. Biol. 2004;49:3817–30.
14. Cucinotta FA, Alp M, Sulzman FM, Wang M. Space radiation risks to the central nervous system. Life Sci. Space Res. 2014;2:54–69.
15. Barcellos-Hoff MH, Brooks AL. Extracellular signaling via the microenvironment: a hypothesis relating carcinogenesis, bystander effects and genomic instability. Radiat. Res. 2001;156:618–27.
16. Heneka MT, O'Banion MK. Inflammatory processes in Alzheimer's disease. J. Neuroimmunol. 2007;184:69–91.

Chapter 2
Biomedical Consequences of Exposure to Space Radiation

The biological effects of exposure to space radiation can be divided into *acute* and *chronic*. Acute effects are the result of exposure to high radiation doses, which may be caused by solar particle events (SPEs), whereas chronic effects are caused by extended exposure to space radiation. Potential effects of either type of exposure include *direct* and *indirect* damage to genetic material, biochemical alterations of cells and/or tissues, carcinogenesis, degenerative tissue effects and cataracts. The extent of these effects is determined by the type of radiation, its flux and the energy spectrum, factors that are incompletely understood.

Other factors that determine radiation damage include age at exposure, gender, and susceptibility to radiation. The quantitative physiological effects of radiation are also poorly understood, due partly to misinterpretation of the exact mechanisms and processes that concern DNA repair. For example, experiments in the 2010s revealed a number of uncertainties that apply to the quality factors used in radiation protection; some studies indicated that physiological damage caused by a specific exposure is only half that previously estimated, a finding explained by the fact that low energy protons inflict more damage than high energy ones. Why? Very simply because the low energy protons take longer to pass through the body and therefore have more time to interact with the tissues.

Yet another poorly understood mechanism is that of Relative Biological Effectiveness (RBE), which is determined to a large degree by radiation type and kinetic energy. How RBE correlates with tumor type or cancer progression is practically unknown because most of the limited experimental data has been conducted on mice, and it is very difficult to extrapolate and apply mice data to humans. It is even more of a challenge to use that data to estimate health risks for cancer, cataracts (Figs. 2.1 and 2.2), and CNS risks. To gain even a cursory insight into the problem will require many, many, *many* more astronauts conducting one-year (or longer) increments followed by post-mission observation times of at least 10 years—at least. Given that the ISS is due to be retired in 2028, this goal is impossible. And even it was, extrapolating data from the ISS to deep space is extremely limited at best in terms of making accurate risk predictions for those who eventually

E. Seedhouse, *Space Radiation and Astronaut Safety*, SpringerBriefs in Space Development, https://doi.org/10.1007/978-3-319-74615-9_2

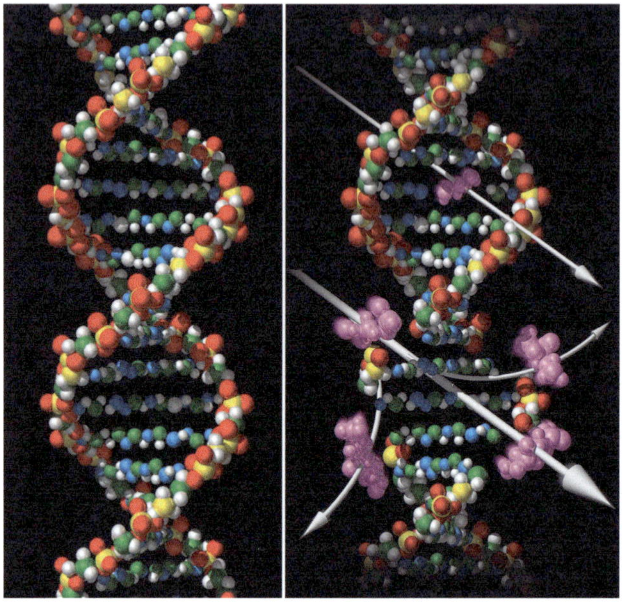

Fig. 2.1 Radiation can cause single and double strand breaks in DNA. (Image courtesy of NASA)

venture beyond Earth orbit. In short, there are myriad knowledge gaps regarding the potential acute and late biomedical risks from galactic cosmic rays (GCRs) and SPEs, but what follows is some of what we know.

Central Nervous System Effects

The potential acute and late risks to the central nervous system (CNS) from GCRs and SPEs have not been a major consideration for crews on board the ISS because these astronauts are exposed to relatively low to moderate doses of ionizing radiation compared to deep space doses (see Chap. 1 for a discussion of the radiation measured en-route to Mars).

> "I'm having these light flashes. I'm seeing this, like, light flashing in my eyeballs. It was like fireworks in your eyeballs. It was spectacular."
> —Charles Duke, Lunar Module Pilot, Apollo 16

The risk presented by GCRs was evident during the Apollo era [1] when astronauts reported the 'light flash' phenomenon described above and caused by HZE nuclei traversing through the retina (Fig. 2.2). As these nuclei traverse, they cause a microlesion. In addition to microlesions in the eye, exposure to HZE nuclei, at doses similar to ones that astronauts will be exposed to during a Mars mission, causes neurocognitive deficits, such as operant reactions. The extent of such a performance

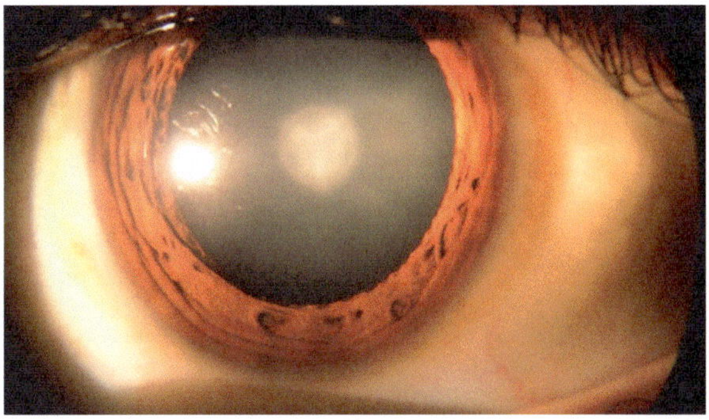

Fig. 2.2 Radiation-induced cataracts are one consequence of spending time in deep space. We know this from the experience of Apollo astronauts, many of whom suffered cataracts within a few years of returning from their mission. There are also several documented cases of astronauts who flew high altitude shuttle missions who suffered from cataracts. (Image courtesy of NASA and Sciencemag.com)

deficit is determined to a large degree by linear energy transfer (LET)[1] and age at the exposure. Other CNS risks include detriments in short-term memory and altered motor function. These are discussed briefly here.

Radiation protection for astronauts is based on a risk of a 1 in 33 probability of death by cancer caused by occupational exposure, a limit that compares with a 1 in 270 risk of loss of crew caused by a flight failure. This risk estimate is determined by human epidemiology combined with quality factors, risk projection models and experimental models, but this method cannot be applied to estimate CNS risks in deep space [2, 3]. The reason for this is that there have been very few humans that have ventured beyond Earth orbit, which means human scaling is next to impossible. What is known is that GCR comprise of protons, helium nuclei and nuclei that travel with high charge and great energy—also known as HZE nuclei. The energy of these nuclei may range from tens of millions of electron volts (MeV)/u to more than 10,000 MeV/u.[2] But these energies do not tell the whole story, because secondary particles are generated as the nuclei pass through shielding and tissue. And since GCR (Fig. 2.3) nuclei have such high energies, they are capable of passing through hundreds of centimeters of material. To get a better understanding of the GCR environment on a trip to Mars, NASA flew the radiation assessment detector (RAD) on the Mars Science Laboratory (MSL) [4].

Table 2.1 shows the GCR dose equivalent measured by the RAD/MSL study. By comparison, astronauts spending 6 months on the ISS are exposed to 80 mSv, or less than a quarter of the exposure of astronauts spending more than 6 months on a trip

[1] LET is the retarding force acting on a charged ionizing particle as it passes through material, whether that material happens to be a spacecraft or an astronaut. The term describes how much energy the particle transfers to the material traversed per unit distance. LET also depends on the type of radiation and the material traversed.

[2] MeV is short for *megaelectron volt* and is equivalent to one million electron volts (eV). One eV is the amount of kinetic energy gained by an electron as it accelerates through an electric potential difference of one volt.

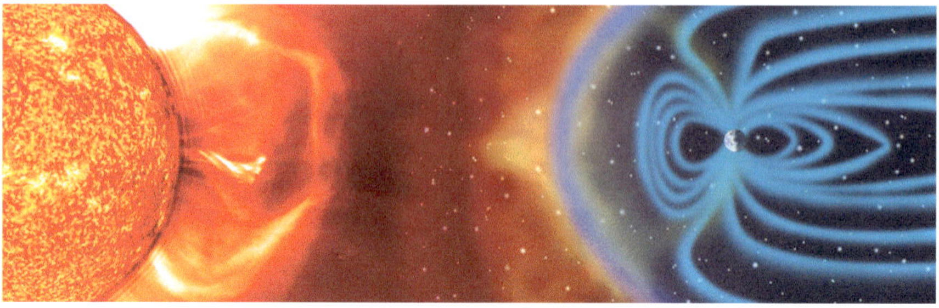

Fig. 2.3 The GCR environment in deep space. (Image courtesy of NASA)

Table 2.1 MSL measurements for average dose-rate on cruise phase to Mars and on Mars surface [4]

	GCR dose equivalent rate (mSv/day)
RAD cruise to Mars	1.84 ± 0.33
RAD Mars surface	0.70 ± 0.17

to Mars [4]. The reason for the difference is that astronauts in LEO are partly protected by Earth's magnetic field, which repels GCR nuclei with energies below 1000 MeV/u. With astronauts taking such a radiation hit it will be necessary to find out as much as possible about the biomedical effects of exposure to GCR. One option is to use non-human primates (NHP) since NHPs and humans share a number of physiological and neurobiological characteristics [5].

For example, NHPs are used to investigate infectious diseases, Alzheimer's, and strokes. Rodents can also be used, although the number of cross-species differences makes clinical determination of CNS health risks. For example, many of the cognitive deficits are known to originate in the frontal cortex in humans, but this area is underdeveloped in rodents. Another example is the difference in risk assessment; death for 50% of a population from a fixed level of radiation exposure occurs at a lower level of exposure for humans than it does for certain strains of rodents. But rodents are the more favored test subject when it comes to radiation research because NHP studies require much higher costs and more thorough ethical review (Table 2.2).

Behavioral Studies of CNS Risks

One radiation-related topic that has received media attention in recent years is the suggestion that exposure to deep space radiation may cause cognitive deficits. One early study that investigated this [7] exposed rats to 1000 MeV/u before testing their spatial memory in a radial maze. In this study, the exposed rats committed more errors than control rats and were unable to develop a spatial strategy to make their way through the maze. A similar study [8] examined mice that had been subjected to two weeks of whole-body irradiation. The results of this study revealed impaired novel object recognition and reduced spatial memory. Another study exposed Wistar

Table 2.2 Disadvantages of rodent models and advantages of NHP comparisons for human radiation risk assessment [6]

(a) Disadvantages of using rodents for investigating biological mechanisms of human radiation risk assessment
• Drastically reduced numbers of neuron cells and synapses compared to humans or NHPs
• Much higher levels of neurogenesis compared to humans or NHPs
• Reduced axon size and number of synapses compared to humans and NHPs
• Reduced pre-frontal cortex size and drastically reduced development of frontal lobe where higher cognition occurs (neocortex is 27% in rats, 72% in monkeys, 80% in humans)
• Performance and cognition tests limited compared to human abilities including inability to respond to visual cues
• Manifestations of pathology of Alzheimer's disease and other late effects extremely limited
• Genetic overlap of the rodent CNS with human CNS is small (20 vs. >80% in NHP)
(b) Advantages of using NHPs to estimate GCR risks
• High level of genetic overlap
• Developed neocortex, including more similar levels of neurons, synapses, and rates of neurogenesis
• NHPs have complex visual behavior, which is not found in rodents
• NHPs have been shown to contain mirror neurons in the pre-frontal cortex that are specific to a particular action, and part of a brain network. Mirror neurons are not believed to exist in rodents
• NHPs develop plaques/pathological lesions observed in Alzheimer's
• NHP share many common features related to memory and cognition with humans that are not found in rodents
• NHPs can be trained for identical cognitive testing including responding to visual cues that are used in human studies making translation of results immediately analyzable

rats [9] to 1000 MeV/u and tested the rats 3 months after exposure. The test in this study was an attentional set shifting task (AST)[3] which only 17% of the irradiated rats were able to complete compared with 78% of control rats.

Altered Neurogenesis[4]

Research has revealed that neurogenesis may be sensitive to radiation [10], which in turn may result in cognitive deficits such as memory [11]. Furthermore, studies have indicated that exposure to high doses of radiation may not only inhibit the generation

[3] The AST measures attention and cognitive flexibility in rats. It is based on the intradimensional/extradimensional component of the Cambridge Neuropsychological Test Automated Battery (CANTAB), which is used to assess cognitive dysfunction in humans. An attentional set is created when a person learns that a set of rules can be used to distinguish between relevant and irrelevant cues.

[4] Neurogenesis is a term that describes the formation of neurons. In adults, this process occurs in the subventricular zone (SVZ) and the subgranular zone of the brain. Scientists are still researching the role that neurogenesis plays in cognition. Since the formation of neurons may be sensitive to radiation, it is possible that long-term exposure may result in cognitive deficits.

of neuronal progenitor cells but that those cells that are generated may not be fully functional [12]. These studies were conducted on mice using doses of 1000 MeV/u [13], which provides some insight into the mechanism of radiation-induced cognitive injury, but for reasons indicated earlier, scaling these results to humans is limited.

Oxidative Damage

Oxidative stress[5] is thought to be implicated in Alzheimer's disease, heart failure, and chronic fatigue syndrome. Since radiation has been shown to increase oxidative damage [14, 15], oxidative stress represents yet another mechanism of radiation-induced cognitive injury. Since antioxidants prevents such damage under normal conditions it would seem logical to suggest that astronauts eat food that contain high levels of antioxidants. For example, a diet high in blueberries and strawberries should help offset oxidative stress. Or melatonin perhaps? Melatonin has high anti-oxidant properties, and studies have shown that it inhibits neurogenesis [16]. The problem with the research to date is that studies have used high dose rates and the biological effects of radiation are different at low dose-rates. Furthermore, studies that have investigated the supposed beneficial effects of antioxidants [17] found no evidence that supplementation actually works. In fact, some studies revealed that antioxidants such as vitamin A, vitamin E and β-carotene might actually be more damaging because taking extra amounts of antioxidants would help the body rescue cells that had been damaged by radiation and that this might alter DNA repair:

> *This study shows for the first time that exposure to radiation levels equivalent to a mission to Mars could produce cognitive problems and speed up changes in the brain that are associated with Alzheimer's disease. These findings clearly suggest that exposure to radiation in space has the potential to accelerate the development of Alzheimer's disease. This is yet another factor that NASA, which is clearly concerned about the health risks to its astronauts, will need to take into account as it plans future missions.*
>
> —Dr. Kerry O'Banion, University of Rochester Medical Center

Alzheimer's Disease

Alzheimer's disease is a neurodegenerative disease that causes dementia in most cases. Common symptoms include short-term memory loss, language problems, disorientation, lack of motivation, and behavioral problems (Fig. 2.4). The disease, which is chronic, begins slowly and symptoms become worse with time. The cause

[5] Oxidative stress is a term that describes the imbalance between the production of free radicals and the ability of the body to neutralize these free radicals through the use of antioxidants. Free radicals are molecules that contain oxygen. These molecules have one or more unpaired electrons, which means they are very reactive with other molecules which in turn means they are capable of chemically interacting with and destabilizing cells such as DNA. Under normal conditions, antioxidants prevent these reactions.

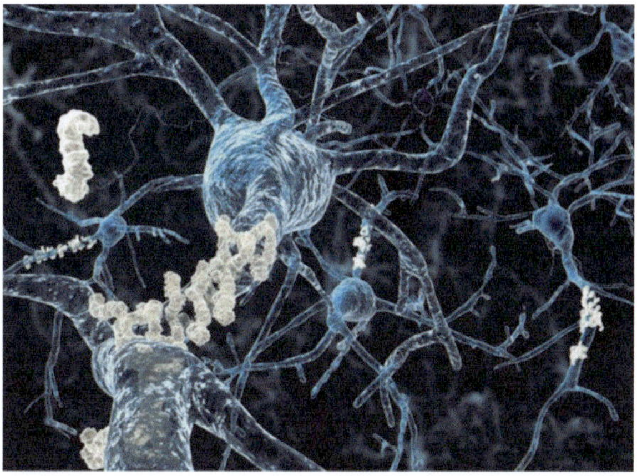

Fig. 2.4 Research has revealed that long-term exposure to the deep space radiation environment may lead to delayed onset of Alzheimer's symptoms. (Image courtesy of Collective Evolution)

of the disease is not completely understood, but the majority of the risk is believed to be genetic. There are no treatments that can stop the disease or even slow its progression. It has been shown in mice that exposure to radiation accelerates the onset of age-related neuronal dysfunction that results in symptoms similar to those exhibited by those suffering from Alzheimer's [18–20].

Some of the most popularized results showing the link between radiation exposure and Alzheimer's pathologies have been those conducted by Dr. O'Banion's laboratory [21]. In one study mice were that were subjected to 1000 MeV/u for 6 months showed decreased cognitive ability [21]: As quoted earlier, "Because iron particles pack a bigger wallop, it is extremely difficult from an engineering perspective to effectively shield against them. One would have to essentially wrap a spacecraft in a 6-foot block of lead or concrete."

Radiation-Induced Bone Loss

Another process affected by radiation is bone remodeling (boxed text). In zero gravity or reduced gravity, bone remodeling is disrupted, resulting in a loss of bone mineral density (BMD). Astronauts on board the International Space Station (ISS) typically lose between 1.0 and 1.2% of BMD per month (Fig. 2.5). This equates to an overall loss of more than 7% during a typical 6-month increment, which in turn results in a two- to threefold increase in fracture risk. For astronauts in low Earth orbit (LEO), this rate of BMD loss is fairly predictable, but beyond LEO, the radiation environment is much harsher and the rate of BMD loss less predictable. This is because bone is damaged more by higher doses of radiation, a fact long documented

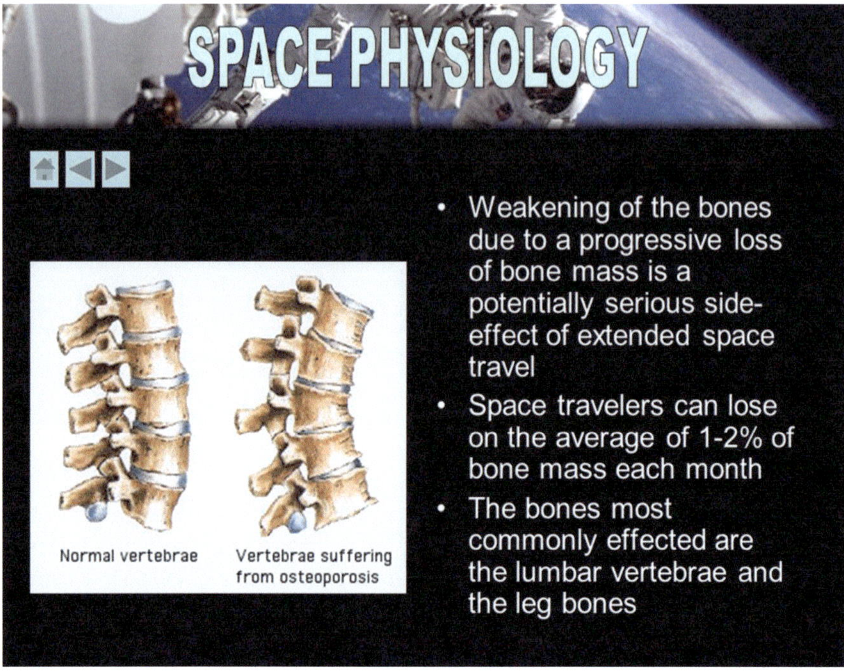

Fig. 2.5 As the slide explains, bone loss is a major occupational hazard among spacefarers, but being exposed to high levels of radiation makes the situation worse because radiation exposure accelerates bone loss. (Image courtesy of NASA)

by the persistent decline in bone volume following exposure to therapeutic radiation in cancer patients [22–24].

Bone Remodeling

On Earth, bone undergoes a process of constant remodeling thanks to the actions of two types of bone cells. The osteoclasts break down bone and the osteoblasts build up bone. This process is continuous thanks to the trigger of constant loading that occurs when walking or running around in a one G environment. But when astronauts enter zero gravity, BMD is lost due to skeletal unloading. This mechanism is characterized by an increase in bone resorption and a decrease in bone formation.

Compounding the effect of osteoradionecrosis is the effect that radiation has on fracture sites [25]. To better understand this it is necessary to familiarize ourselves with the biological damage caused by radiation. Remember, there are two types of

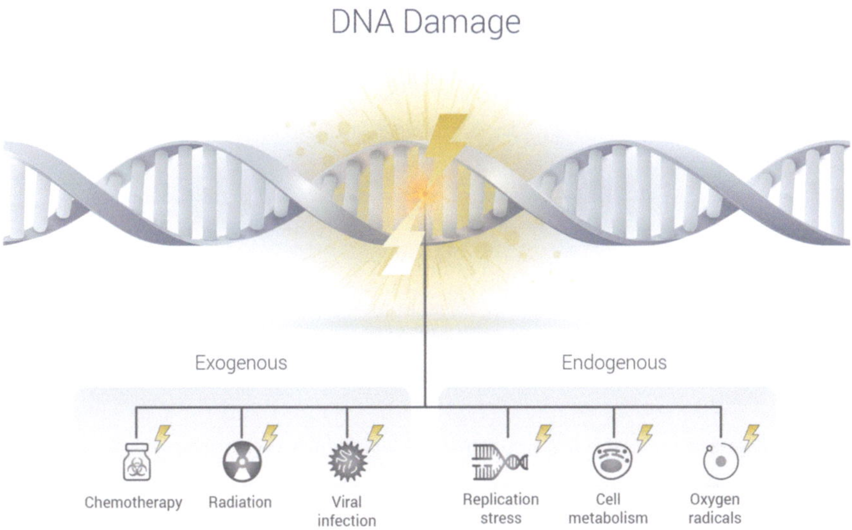

Fig. 2.6 The body has outstanding repair mechanisms, but it can only do so much. After each radiation strike, repairs become less effective. (Image courtesy of NASA)

radiation—*non-ionizing* and *ionizing*. Non-ionizing radiation does not cause significant biological damage because it does not displace electrons from an atom. Ionizing radiation on the other hand has sufficient energy to displace electrons from an atom, thereby creating an ionization event. The energy of this ionization event can break molecular bonds and thereby cause biological damage such as single strand or double strand breaks in DNA (Fig. 2.6). Although the body is tremendously resilient in its capacity to repair radiation damage, some cells ultimately die in this onslaught of ionizing radiation. Worse, some cells may actually propagate the ionized-induced damage to progeny.

We already know that the space radiation environment comprises a mix of ions generated by SPEs and GCR. We also know that, thanks to the data sent back from the RAD strapped onto the Curiosity rover, the transit to Mars will result in a radiation exposure of more than 3 Sieverts. This means that tissue-dose rates from space radiation will be about 1–2.5 mSv per day (the *annual* terrestrial dose, incidentally), but solar flare dose-rates may increase this number to more than 100 mSv/day, even if inside a shielded vehicle. Furthermore, an astronaut conducting a deep space EVA during a solar flare event may be exposed to a dose rate as high as 250 mSv per day. By comparison, cancer patients receive daily dose (fractions) of about 6 Sv targeted at the tumor, but these doses are delivered over a period of minutes [26, 27].

We have a fairly good understanding of these processes because ionizing radiation has long been used as a treatment for malignancies (Fig. 2.7) and has been a factor in reducing cancer mortality. One of the main reasons bone mineral density is reduced following irradiation is because osteoblasts and osteoclasts are damaged.

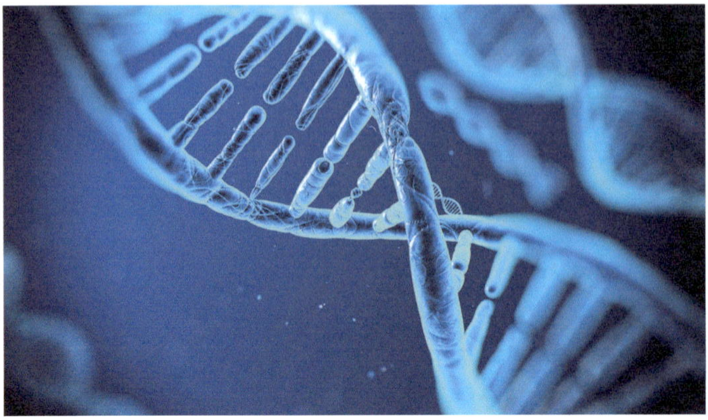

Fig. 2.7 If genetic material is damaged repeatedly, tumors may occur. (Image courtesy of Toma Biosciences)

The osteoblasts and osteoclasts are two bone cells that work together to remodel bone; the osteoclasts break down bone and the osteoblasts build up bone. But when these bone cells are damaged, bone formation is impaired due to cell-cycle arrest. One of the processes by which the osteoclasts and osteoblasts are damaged is by oxidative stress caused by the radiation since it is this oxidative stress that damages osteoprogenitors. To begin with, irradiation causes an increase in osteoclast number, which thereby causes osteoporosis [28]. Shortly after exposure there is a decline in the number of osteoclasts and osteoblasts, which results in suppression of bone remodeling and degradation of bone quality.

However, the side effects of this treatment have concerned oncologists and will be of concern to flight surgeons responsible for the health of astronauts embarked on exploration class missions (ECM) beyond LEO. This is because bones within the irradiated area are at a much higher fracture risk [29, 30]. For example, patients undergoing breast cancer treatment may have rib fracture rates that exceed 15%. This is of concern to astronauts on long duration missions because their bones will already be weakened due to the loss of BMD simply as a consequence of being in microgravity. Thus, these astronauts may be at high risk of traumatic and/or spontaneous fracture [31]. But what levels of radiation exposure cause these problems? Well, studies using ECM-relevant radiation [30] have revealed that whole-body exposure to 2 Gy can cause a reduction in trabecular bone quality, while bone loss can persist for as long as 4 months following exposure to irradiation as low as 1 Gy. Similar studies also revealed that functional bone loss can be detected after just three days following exposure to 2 Gy.

As is evident from these research findings, the effects of radiation result in a cascade of changes. In addition to the reduction in bone mineral density and the impact on healing, the way in which bone is weakened will be of concern to flight surgeons tasked with keeping ECM crewmembers fracture-free. For example, the loss of trabecular bone means cortical bone must now deal with a greater proportion

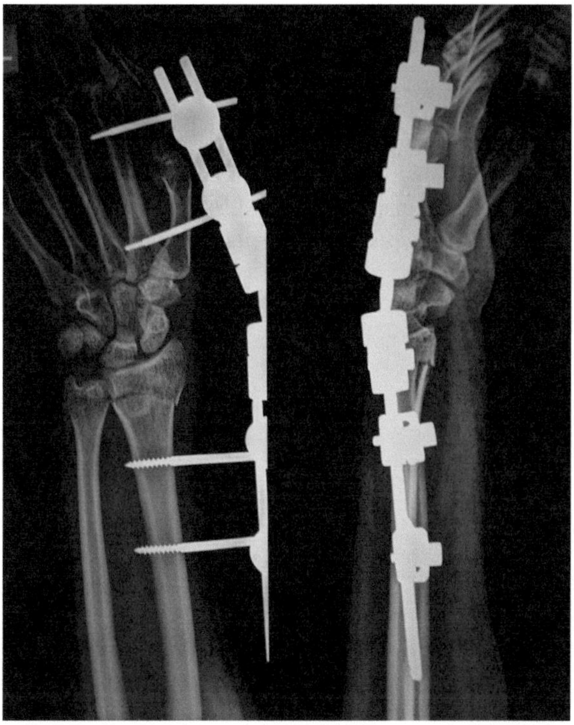

Fig. 2.8 External fixation. The means to treat even simple fractures will probably exceed the medical capabilities of exploration class spacecraft. (Image courtesy of Ashish j29)

of loads on the skeleton. This in turn means that cortical bone will be increasingly less able to resist the torsional and bending loads, a change that may be exacerbated by any defect in the bone such as a porous hole. The net effect of all these radiation-induced effects is an overall disruption of load distribution that results in a compromised structural integrity of the bone (Fig. 2.8). For astronauts about to land on Mars, this is not an optimum situation. It is important to remember that after their ISS increments, astronauts return to Earth with increased fracture risk but that this doesn't mean a fracture is imminent. It just means that due to the loss of BMD, there is a greater chance of fracture after their return. But this increased fracture risk is dramatically reduced thanks to the rehabilitation schedule that astronauts follow after their return to Earth, which results in regeneration of bone. On Mars there will be limited regeneration.

Osteitis and Osteoradionecrosis

Radiation-induced bone loss is termed *osteitis,* whereas *osteoradionecrosis* is a condition in which the bone's ability to withstand trauma is reduced. In this condition, non-healing bone may be susceptible to infection, and the ability of the bone to heal is further complicated by hypovascularization. Basically, as the body is subjected to more and more radiation, the very small blood vessels inside the bone are destroyed. This is devastating because it is these blood vessels that carry nutrients and oxygen to the bone. Without blood vessels to do this, the bone simply dies. On Earth, one treatment option for patients with osteoradionecrosis is hyperbaric oxygen therapy (although even with this treatment, less than 30% of patients will survive), but this will not be an option on an interplanetary spaceship.

Weakened bone. Increased fracture risk. Radiation-induced bone loss. These are all critical mission concerns that will require flight surgeons to include 'bone complications' as part of the exploration mission treatment planning process. For example, flight surgeons will need to consider at-risk bone volumes as a critical concern and plan accordingly. Likewise, the increased suppression of bone formation leading to compromised bone strength will require a rethinking of the medical capabilities on board the Mars-bound spacecraft. An external fixation kit perhaps? Hopefully that won't be required if research can more accurately elucidate the dose thresholds for deep space missions. This will require studies that consider the effects of a range of GCR species and energies across the timeframe of an exploration mission.

References

1. Pinsky LS, Osborne WZ, Bailey JV, Benson RE, Thompson LF. Light flashes observed by astronauts on Apollo 11 through Apollo 17. Science. 1974;183(4128):957–9.
2. Cucinotta FA, Wang H, Huff JL. Risk of acute or late central nervous system effects from radiation exposure. Human Health and Performance Risks of Space Exploration Missions NASA SP 2009, 2009, p. 345. Chapter 6.
3. Curtis SB, Vazquez ME, Wilson JW, Atwell W, Kim M, Capala J. Cosmic ray hit frequencies in critical sites in the central nervous system. Adv. Space Res. 1998;22(2):197–207.
4. Zeitlin CJ, Hassler DM, Cucinotta FA, Ehresmann B, Wimmer-Schweingruber RF, et al. Measurements of energetic particle radiation in transit to Mars on the Mars Science Laboratory. Science. 2013;340:1080–4.
5. Weatherall D. The use of non-human primates in research. Sponsored by the UK Medical Research Council, Welcome Trust, and the Royal Society, 2006.
6. Singh VK, Newman VL, Berg AN, MacVittie TJ. Animal models for acute radiation syndrome drug discovery. Expert Opin. Drug Discovery. 2015;10(5):497–517.
7. Denisova NA, Shukitt-Hale B, Rabin BM, Joseph JA. Brain signaling and behavioral responses induced by exposure to (56)Fe-particle radiation. Radiat. Res. 2002;158:725–34.
8. Haley GE, Yeiser L, Olsen RH, Davis MJ, Johnson LA, Raber J. Early effects of whole-body ^{56}Fe irradiation on hippocampal function in C57BL/6J mice. Radiat. Res. 2013;179(5):590–6.
9. Britten RA, Davis LK, Johnson AM, Keeney S, Siegel A, Sanford LD, Singletary SJ, Lonart G. Low (20 cGy) doses of 1 GeV/u ^{56}Fe-particle radiation lead to a persistent reduction in the spatial learning ability of rats. Radiat. Res. 2012;177(2):146–51.
10. Mizumatsu S, Monje ML, Morhardt DR, Rola R, Palmer TD, Fike JR. Extreme sensitivity of adult neurogenesis to low doses of X-irradiation. Cancer Res. 2003;63(14):4021–7.
11. Squire LR. Memory and the hippocampus: a synthesis from findings with rats, monkeys, and humans. Psychol. Rev. 1992;99(2):95–231.
12. Casadesus G, Shukitt-Hale B, Stellwagen HM, Smith MA, Rabin BM, Joseph JA. Hippocampal neurogenesis and PSA-NCAM expression following exposure to ^{56}Fe particles mimics that seen during aging in rats. Exp. Gerontol. 2005;40:249–54.

13. Riviera PD, Shih H, LeBlanc JA, Cole MG, Amaral AZ, Mukherjee S, et al. Acute and fractionated exposure to high-let ^{56}Fe HZE-particle radiation both result in similar long-term deficits in adult hippocampal neurogenesis. Radiat. Res. 2014;180:658–67.
14. Limoli CL, Giedzinski E, Rola R, Otsuka S, Palmer TD, Fike JR. Radiation response of neural precursor cells: linking cellular sensitivity to cell cycle checkpoints, apoptosis and oxidative stress. Radiat. Res. 2004;161(1):17–27.
15. Limoli CL, Giedzinski Baure EJ, Rola R, Fike JR. Redox changes induced in hippocampal precursor cells by heavy ion irradiation. Radiat. Environ. Biophys. 2007;46(2):167–72.
16. Manda K, Ueno M, Anzai K. Memory impairment, oxidative damage and apoptosis induced by space radiation: ameliorative potential of alpha-lipoic acid. Behav. Brain Res. 2008;187(2):387–95.
17. Bjelakovic G, Nikolova D, Gluud LL, Simonetti RG, Gluud C. Mortality in randomized trials of antioxidant supplements for primary and secondary prevention: systematic review and meta-analysis. JAMA. 2007;297:842–57.
18. Vlkolinský R, Krucker T, Smith AL, Lamp TC, Nelson GA, Obenaus A. Effects of lipopolysaccharide on ^{56}Fe-particle radiation-induced impairment of synaptic plasticity in the mouse hippocampus. Radiat. Res. 2007;168(5):462–47.
19. Vlkolinský R, Krucker T, Nelson GA, Obenaus A. ^{56}Fe-particle radiation reduces neuronal output and attenuates lipopolysaccharide-induced inhibition of long-term potentiation in the mouse hippocampus. Radiat. Res. 2008;169(5):523–53.
20. Vlkolinsky V, Titova E, Krucker T, Chi BB, Staufenbiel M, Nelson GA, Obenaus A. Exposure to ^{56}Fe-particle radiation accelerates electrophysiological alterations in the hippocampus of APP23 transgenic mice. Radiat. Res. 2010;173(3):342–52.
21. Cherry JD, Liu B, Frost JL, Lemere CA, Williams JP, et al. Galactic cosmic radiation leads to cognitive impairment and increased Aβ plaque accumulation in a mouse model of Alzheimer's disease. PLoS ONE. 2012;7(12):e53275.
22. Lloyd SA, Bandstra ER, Travis ND, Nelson GA, Bourland JD, Pecaut MJ, Gridley DS, Willey JS, Bateman TA. Spaceflight-relevant types of ionizing radiation and cortical bone: potential LET effect? Adv. Space Res. 2008;42:1889–97.
23. Hopewell JW. Radiation-therapy effects on bone density. Med. Pediatr. Oncol. 2003;41:208–11.
24. Lang T, LeBlanc A, Evans H, Lu Y, Genant H, Yu A. Cortical and trabecular bone mineral loss from the spine and hip in long-duration spaceflight. J. Bone Miner. Res. 2004;19:1006–12.
25. Guise TA. Bone loss and fracture risk associated with cancer therapy. Oncologist. 2006; 11:1121–31.
26. Overgaard M. Spontaneous radiation-induced rib fractures in breast cancer patients treated with postmastectomy irradiation. A clinical radiobiological analysis of the influence of fraction size and dose-response relationships on late bone damage. Acta Oncol. 1988;27:117–22.
27. Rabelo GD, Beletti ME, Dechichi P. Histological analysis of the alterations on cortical bone channels network after radiotherapy: a rabbit study. Microsc. Res. Tech. 2010;73:1015–8.
28. Sakurai T, Sawada Y, Yoshimoto M, Kawai M, Miyakoshi J. Radiation-induced reduction of osteoblast differentiation in C2C12 cells. J. Radiat. Res. 2007;48:515–21.
29. Sugimoto M, Takahashi S, Toguchida J, Kotoura Y, Shibamoto Y, Yamamuro T. Changes in bone after high-dose irradiation. Biomechanics and histomorphology. J. Bone Joint Surg. Br. 1991;73:492–7.
30. Willey JS, Lloyd SA, Robbins ME, Bourland JD, Smith-Sielicki H, Bowman LC, Norrdin RW, Bateman TA. Early increase in osteoclast number in mice after whole-body irradiation with 2 Gy X rays. Radiat. Res. 2008;170:388–92.
31. Green DE, Adler BJ, Chan ME, Rubin CT. Devastation of adult stem cell pools by irradiation precedes collapse of trabecular bone quality and quantity. J Bone Miner Res. 2012;27:749–59.

Chapter 3
Setting Acceptable Risk Levels for Astronauts

The permissible exposure levels (PELs) for radiation exposure that NASA sets for its astronauts are meant to prevent inflight risks and to limit risk to a level that is acceptable from an ethical, moral, and financial standpoint. In the 1960s and 1970s, PELs were set based on recommendations made by the National Academy of Sciences. In the 1980s, more data on radiation exposure had been accumulated, and NASA asked the National Council on Radiation Protection (NCRP) to re-assess the dose limits for astronauts working in LEO. This re-assessment culminated in the NCRP Report No. 98, published in 1989, which recommended age and gender career dose limits that applied a 3% increase in cancer mortality as a risk limit.

NCRP Report No. 98 was followed by revisions to the acceptable level of radiation risk in LEO in NCRP reports published in 1997 and 2000 [1, 2]. The astronaut risk level of a 3%[1] increase risk in cancer over a lifetime is similar to the risk level applied to workers in nuclear facilities. The difference is that the radiation doses that nuclear workers are exposed to correspond to a lifetime of exposure to relatively low radiation doses compared with the exposure limit in LEO. For example, radiation workers in nuclear reactors rarely approach an annual average exposure of 2 mSv. This is significantly below the effective dose of 80 mSv astronauts are exposed to during a six-month increment on board the International Space Station (ISS, Figs. 3.1 and 3.2). The point is that ground workers are exposed to *chronic* exposure at low levels (compared with radiation levels in LEO) of radiation over a long period of time, whereas astronauts are exposed both chronic *and* acute radiation.

[1] There are many approaches to setting acceptable risk levels. One is to set an unlimited risk level, but this would not be popular with astronauts or their families. Another option is to base risk on life-loss from radiation-induced cancer against cancer deaths in the general population. The current method uses the reference point of a ground-based radiation worker.

E. Seedhouse, *Space Radiation and Astronaut Safety*, SpringerBriefs in Space Development, https://doi.org/10.1007/978-3-319-74615-9_3

Fig. 3.1 Astronauts performing spacewalks are particularly susceptible to the effects of radiation. (Image courtesy of NASA)

Fig. 3.2 European Space Agency astronaut Alexander Gerst looking through the cupola. (Image courtesy of NASA)

Permissible Exposure Limits

NASA's PELs take into account a number of factors, including age, gender, latency effects, differences in tissue types and differences in lifespan between genders [3, 4]. When all these factors are considered, a risk projection calculation can be made, and a risk of exposure induced death (REID) from fatal cancer can be made (Table 3.1).

Another limit that NASA applies is one for non-cancer effects. For example, radiation exposure may also cause prodromal effects such as nausea and fatigue, and it may also cause heart disease, dementia, and central nervous system (CNS) damage. For dose limits for non-cancer effects, NASA calculates the relative biological effectiveness (RBE) factor to the major organs of the body as indicated in Tables 3.2 and 3.3.

Other space agencies such as ESA and the Russian Space Agency (Roscosmos) estimate dose limits based on data published by the International Commission on Radiological Protection (ICRP) [5]. Although these dose limits can be applied to LEO workers, there are no operationally approved dose limits for deep space, although estimates have been made for the risk astronauts will be subjected to when traveling to the Moon (Fig. 3.3) or Mars (Table 3.4).

Table 3.1 Theoretical dose limits for one-year missions based on 3% REID[a]

	E (mSv) for 3% REID (average life-loss per death [y])	
Age at exposure	Males	Females
30	620 (15.7)	470 (15.7)
35	720 (15.4)	550 (15.3)
40	800 (15.0)	620 (14.7)
45	950 (14.2)	750 (14.0)
50	1150 (12.5)	920 (13.2)
55	1470 (11.5)	1120 (12.2)

[a]Adapted from: Radiation risk acceptability and limitations. Cucinotta F. (https://three.jsc.nasa.gov/articles/AstronautRadLimitsFC.pdf). 12-21-2010

Table 3.2 Dose limits for short-term and career non-cancer effects[a]

Organ	30-day limit	1 year limit	Career limit
Lens	1000 mGy-Eq[b]	2000 mGy-Eq	4000 mGy-Eq
Skin	1500	3000	6000
Blood-forming organs	250	500	N/A
Heart	250	500	1000
Central nervous system	500	1000	1500

[a]Adapted from: Radiation Health Risk Projections Briefing to NAC HEOMD/SMD Joint Committee April 7, 2015

[b]Milli-Gray Equivalent. The Standard International (SI) unit of absorbed dose is the gray (Gy). One Gy is a measure of the absorption of one joule of radiation energy per kilogram. Note: The gray is different from the Sievert (Sv), which is the SI unit that represents the biological effect of radiation

Table 3.3 Tissue weighting factor calculated by attributing an estimate for a tissue's contribution to cancer

Organ	Tissue weighting factor	Organ	Tissue weighting factor
Gonads	0.20	Liver	0.05
Bone marrow (red)	0.12	Esophagus	0.05
Colon	0.12	Thyroid	0.05
Lung	0.12	Skin	0.01
Stomach	0.12	Bone surface	0.01
Bladder	0.05	Remainder	0.05
Breast	0.05	Adrenals, brain, intestine, kidney, muscle, spleen	

Fig. 3.3 ESA's Lunar Village. (Image courtesy of ESA)

The numbers presented in Table 3.4 are not encouraging for space agencies hoping to send astronauts to the Moon for lengthy stays or to Mars (Fig. 3.4). A lunar stay of 180 days results in an exposure of 170 mSv, which is twice that of a 180-day stay on board the ISS. But a roundtrip to Mars results in an exposure of more than 1000 mSv, which exceeds NASA's guidelines (Tables 3.5 and 3.6), which state that no astronaut should be exposed to more than this amount of radiation in a lifetime. To exceed this limit is to increase the risk of developing a fatal cancer by 5% or more. For astronauts embarking on such a trip, the accumulated radiation dose would be akin to getting a whole-body computer tomography (CT) scan every 5 days [6, 7]. Also, exposure to more than 1000 mSv would not amount to a 'measured dose' (i.e., over a lifetime) because crewmembers would receive this amount in less than three years. Being exposed to such a large amount of radiation in such a short period of time would result in changes at the cellular level and possibly mild acute radiation syndrome symptoms.

Table 3.4 Calculation of %-REID from fatal cancer for lunar or Mars missions at solar minimum behind a 5-g/cm^2 aluminum shield[a]

Mission type	*E* (Sv) %-REID	
Males (40 years old)		
Lunar (180 days)	0.17	0.68
Mars swingby (600 days)	1.03	4.0
Mars exploration (1000 days)	1.07	4.2
Females (40 years old)		
Lunar (180 days)	0.17	0.82
Mars swingby (600 days)	1.03	4.9
Mars exploration (1000 days)	1.07	5.1

The effective dose, *E* is averaged over tissues susceptible to cancer risk
[a]Adapted from Cucinotta and Durante, 2006

Fig. 3.4 Astronauts traveling to Mars will be exposed to radiation levels beyond the career thresholds set by space agencies. (Image courtesy of NASA)

Table 3.5 Career exposure limits for NASA astronauts[a]

Age	25	35	45	55
Male (Sv)	1.50	2.50	3.25	4.00
Females (Sv)	1.00	1.75	2.50	3.00

[a]An astronaut's organ and career exposure limits are determined by age and gender. An average dose for an Earth-bound person is 0.0036 Sv, whereas someone who works in a nuclear power plant may be exposed to as much as 0.05 Sv per year without exceeding international standards. As you can see, the limit for astronauts is significantly higher

Table 3.6 Depth of radiation penetration and exposure limits for astronauts and public (Sv)

	Exposure Interval	Blood-Forming Organs (5-cm depth)	Eyes (0.3-cm depth)	Skin (0.01-cm depth)
Astronauts	30 days	0.25	1.0	1.5
	Annual	0.50	2.0	3.0
	Career	1–4	4.0	6.0
General public	Annual	0.001	0.015	0.05

The ALARA and AHARS Principles

Although NASA may be able to reduce the amount of radiation exposure by extra shielding, another option might be to apply a different risk strategy [8–10]. Traditionally, the agency has applied the maxim of As Low As Reasonably Achievable (ALARA) when it comes to exposing astronauts to radiation. Another option may be to change this principle to As High as Relatively Safe (AHARS).

Since astronauts are exposed to so much radiation, they are classified as radiation workers. But because they are exposed to so much radiation, the amount of radiation that astronauts are exposed to exceeds all terrestrial limits. This is why the Occupational Safety and Health Administration (OSHA) gave NASA a waiver that gave the agency the green light to create its own guidelines, which have been outlined in this chapter. But even these very high limits will be exceeded during a roundtrip to Mars, which means the agency will probably need to grant exceptions, allowing NASA to put their astronauts at even greater risk. That risk may be reduced with the operation of faster propulsion systems such as VASIMR, or it may be reduced with better shielding (see Chap. 6). But until either or both of these technologies becomes available, radiation risk will remain an intractable problem.

References

1. National Council on Radiation Protection and Measurements, NCRP. Uncertainties in Fatal Cancer Risk Estimates Used in Radiation Protection, NCRP Report 126, Bethesda MD, 1997.
2. National Council on Radiation Protection and Measurements, NCRP. Acceptability of Risk from Radiation – Application to Human Spaceflight. Symposium Proceedings. NCRP, Bethesda MD, 1997a. National Council on Radiation Protection and Measurements, Recommendations of Dose Limits for Low Earth Orbit. NCRP Report 132, Bethesda MD, 2000.

3. National Academy of Sciences, NAS. National Academy of Sciences Space Science Board, Report of the Task Group on the Biological Effects of Space Radiation. Radiation Hazards to Crews on Interplanetary Mission National Academy of Sciences, Washington, D.C., 1996.
4. BEIR VII, National Academy of Sciences Committee on the Biological Effects of Radiation, Health Risks from Exposure to Low Levels of Ionizing Radiation. National Academy of Sciences Press, Washington D.C., 2006.
5. ICRP Publication 60, Recommendations of the International Commission on Radiological Protection, Pergamon Press Inc., 1991. ICRP Publication 103 The 2007 Recommendations of the International Commission on Radiological Protection. Annals of the ICRP 37/2-4, 2007. 8. National Academy of Sciences, NAS, National Research Council, Radiation Protection Guides and Constraints for Space-Mission and Vehicle-Design Studies Involving Nuclear System, Washington D.C., 1970.
6. Cucinotta FA, Schimmerling W, Wilson JW, Peterson LE, Saganti P, Badhwar GD, Dicello JF. Space radiation cancer risks and uncertainties for Mars missions. Radiat. Res. 2001;156: 682–8.
7. Cucinotta FA, Kim MH, Willingham V, George KA. Physical and biological organ dosimetry analysis for International Space Station astronauts. Radiat. Res. 2008;170:127–38.
8. National Council on Radiation Protection and Measurements, NCRP. Guidance on Radiation Received in Space Activities. NCRP Report 98, NCRP, Bethesda MD, 1989.
9. National Council on Radiation Protection and Measurements, NCRP. Operational radiation protection in low Earth orbit. NCRP Report 142, Bethesda MD, 2003.
10. National Council on Radiation Protection and Measurements. Information needed to make radiation protection recommendations for space missions beyond low-Earth orbit. NCRP Report No. 153, Bethesda, MD, 2006.

Chapter 4
Space Radiation Analysis Group

The Space Radiation Analysis Group (SRAG), based at Johnson Space Center (JSC) in Houston, is responsible for ensuring that astronauts are not exposed to radiation levels above NASA's established limits. To perform this function, the group of health physicists, programmers and contractors provides:

1. Preflight astronaut exposure projections based on mission parameters.
2. Real-time radiation protection support.
3. Inflight radiation projections for extravehicular activity.
4. Assessments of radiation-producing equipment carried on board spacecraft.
5. Radiation monitoring instruments (Figs. 4.1 and 4.2) to measure radiation exposure and characterize the radiation environment inside and outside the spacecraft.
6. Crew radiation exposure modeling capabilities.

To perform these tasks, the SRAG team consults the following publications[1]:

- Radiation Exposures in Space and the Potential of Central Nervous System Effects—Committee SC-1-24 Published: NCRP Commentary No. 23 (2014)
- Radiation Protection for Space Activities: Supplement to Previous Recommendations. NCRP Report No. 167 (2010)
- Potential Impact of Individual Genetic Susceptibility and Previous Radiation Exposure on Radiation Risk for Astronauts, NCRP Report No. 153 (2006)
- Information Needed to Make Radiation Protection Recommendations for Space Missions Beyond Low-Earth Orbit NCRP Report No. 142 (2002)
- Operational Radiation Safety Program for Astronauts in Low Earth Orbit: A Basic Framework, NCRP Report No. 132 (2000)—Current Basis of NASA Std 3001 Limits Radiation Protection Guidance for Activities in Low-Earth Orbit

[1] Briefing to NAC HEOMD/SMD Joint Committee April 7, 2015.

E. Seedhouse, *Space Radiation and Astronaut Safety*, SpringerBriefs in Space Development, https://doi.org/10.1007/978-3-319-74615-9_4

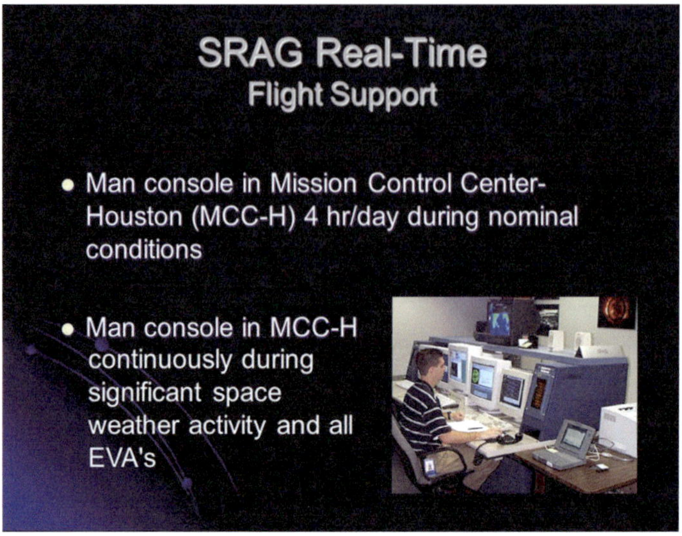

Fig. 4.1 The Space Radiation Assessment Group monitors and tracks radiation exposure on board the ISS. (Image courtesy of NASA)

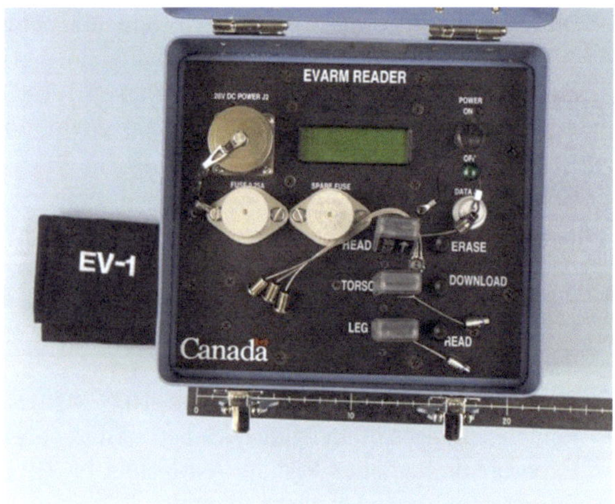

Fig. 4.2 SRAG (Image courtesy of NASA)

SRAG Console

SRAG's console is located in the Mission Control Center, Houston. The personnel manning the console are responsible for:

1. Analyzing space weather (Fig. 4.3) data and forecasts.
2. Interfacing with NOAA to determine big picture space weather conditions.
3. Checking spacecraft status and crew schedules for scheduled and unscheduled EVAs.

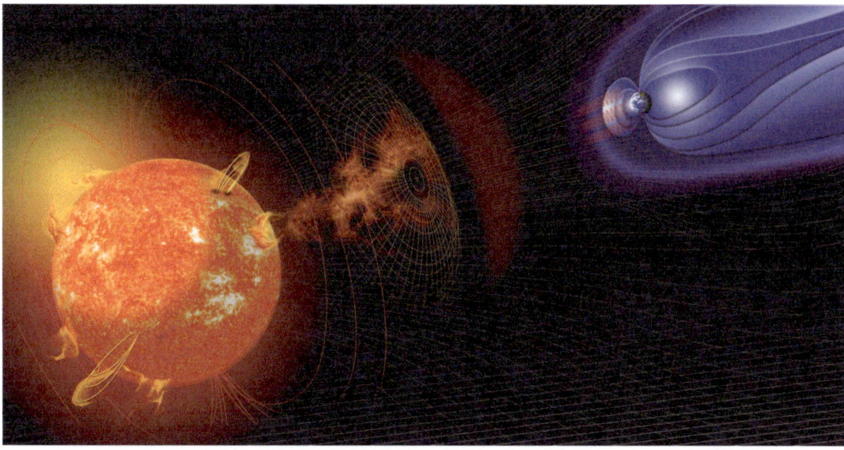

Fig. 4.3 The Canadian Space Agency's EVARM radiation monitor. (Image courtesy of CSA)

Fig. 4.4 Yihua Zheng and Antti Pulkkinen monitor space weather for Goddard's Space Weather Center. (Image courtesy of NASA)

4. Reporting crew radiation exposure status and space weather forecasts to the flight director.
5. Providing regular cumulative crew radiation exposure updates to the flight director.
6. Planning for contingency EVA support (Fig. 4.4).
7. Interfacing with flight controllers the day before scheduled EVAs.
8. Providing EVA radiation exposure analysis to Flight Surgeon.

9. Providing Go/No Go recommendation for EVA.
10. Monitoring space radiation in real-time.
11. Receiving immediate notifications from NOAA of solar particle events (SPEs).
12. Warning flight management of changes to space weather that may affect EVA operations.
13. Analyzing radiation environment and make recommendations for continuing, delaying, or terminating EVAs.
14. Monitoring ISS radiation data.

SRAG Interfaces

To achieve these tasks, the SRAG interfaces with several entities as outlined in Fig. 4.5 below.

Via these interfaces, SRAG is able to develop a complete understanding of the following key parameters that affect astronaut exposure to radiation:

1. Structure of the spacecraft
2. Materials used to construct the vehicle
3. Altitude and inclination of the spacecraft
4. Status of outer zone electron belts
5. The interplanetary proton flux

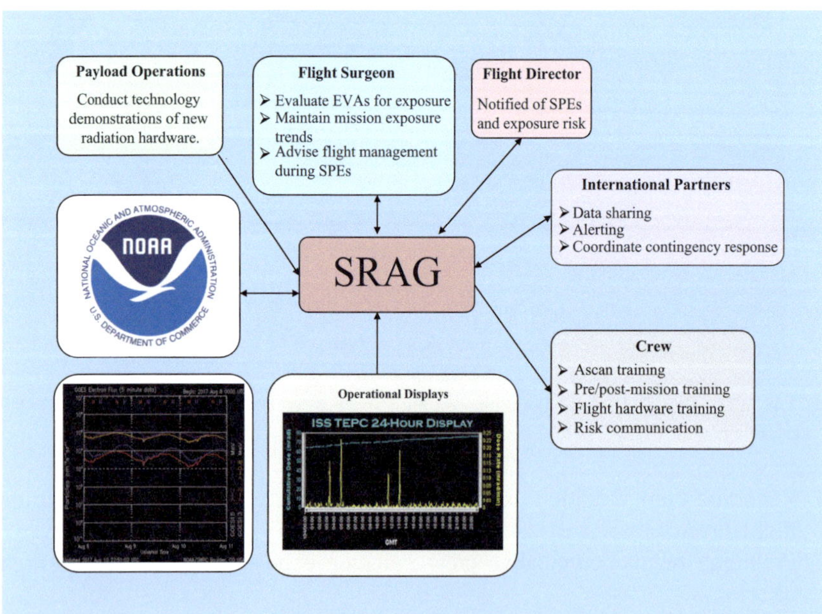

Fig. 4.5 SRAG interfaces. (Image courtesy of NASA)

6. Geomagnetic field conditions
7. Solar cycle position
8. EVA start time and duration.

Radiological Support

To provide radiological support, SRAG mans the radiation consoles in the MCC 4 h every day during nominal space weather conditions, continuously during significant weather activity, and whenever astronauts conduct spacewalks. Data received from NOAA's space weather satellites and the Space Environment Center (SEC) is monitored and examined by SRAG team members to detect trends that might lead to significant space weather and also to develop better forecasting of such weather. If significant space weather *does* occur, the SRAG team may recommend rescheduling EVAs and, if the weather looks like it might lead to significant increased radiation exposure levels, the team may recommend that astronauts seek the best shielded areas in the ISS.

To support mission planning, which is another of the SRAG team's tasks, the team utilizes a suite of tools that include time-resolved models of Earth's magnetic field and trajectory propagator algorithms. Data from these and other tools are integrated with mission data such as the inclination of the spacecraft and the timing of spacewalks to accurately characterize crew radiation exposure. The SRAG team's mandate also extends to any equipment that produces radiation, so any investigator who wants to fly a radioactive isotope on board the ISS must first meet the SRAG team's review. This review assesses the internal and external dose rates astronauts might receive based on their likely proximity to the isotope and also reviews proposed containment and decontamination procedures.

Radiation Instruments

SRAG has access to a suite of instruments on board the ISS that monitor the radiation environment, one of which is the Tissue Equivalent Proportional Counter (TEPC).

Tissue Equivalent Proportional Counter

The TEPC (Fig. 4.6) is an automatic dosimetry system that is comprised of a spectrometer and a detector unit, which together collect a record of the ISS radiation environment and develops exposure history records for each astronaut [1, 2]. The detector unit, which is omni-directional, is surrounded by tissue equivalent plastic and propane gas, which provides an energy deposition that is almost the same as human tissue, or more specifically an approximation of cell wall and cell body. If the crew wants to read their current radiation exposure level or their incremental

Fig. 4.6 The ISS IV-TEPC. (Image courtesy of NASA)

dose, all they need to do is read the display. When the TEPC is powered on it is in either 'standby' or 'acquire' mode. In the 'standby' mode, the TEPC can download data (which is sent to SRAG personnel), whereas in the 'acquire' mode the TEPC is collecting data. The unit also has an alarm capability that informs the crew and the ground if predetermined radiation levels are exceeded.

Charged Particle Directional Spectrometer

The Charged Particle Directional Spectrometer (CPDS) measures the charge, energy, and direction of a radiation particle as it passes through the detectors embedded in the instrument [3, 4]. Inside the CPDS (see also Chap. 5) are 13 detectors arranged in a stack as follows:

1. *A-1 and A-2 detectors*. There are three A-1 and three A-2 1-mm thick silicon detectors and their function is to record the passage of a charged particle through the CPDS.
2. *Position sensitive detectors (PSDs)*. There are three of these 0.3-mm thick silicon detectors arranged in 24 horizontal and 24 vertical 1-mm strips. This arrangement is designed so the unit can track and record the x and y coordinates of a radiation particle as it travels along the z axis.
3. *B-detectors*. There are three of these 5-mm thick lithium-silicon detectors.
4. *Cerenkov or C-detector*. This detector comprises a 1-cm thick crystal sapphire detector combined with a photo-multiplier tube. The photo-multiplier tube gathers light, which in turn provides information about particle velocity.

Perhaps the easiest way to understand how the CPDS functions is to think of it as a telescope. Four of these units are used on board the ISS, each with specific functions as follows:

- (1 unit) Intra-vehicular Charged Particle Directional Spectrometer (IV-CPDS) is used inside the ISS to display real-time calculations of dose rate for crewmembers.
- (3 units) Extra-vehicular Charged Particle Directional Spectrometer (EV-CPDS). These are used outside the ISS for the purpose of characterizing the radiation environment during spacewalks [5, 6].

Radiation Area Monitor

The Radiation Area Monitor (RAM) comprises a group of thermoluminescent detectors (TLDs) surrounded by Lexan, a material that responds to radiation by electronic excitation of the TLD materials [7, 8]. The monitor is just one of many prepared by the SRAG team. After return of the monitors, SRAG personnel working in the Radiation Dosimetry Lab analyze the data before submitting reports to the Flight Surgeon and the Space Radiation Health Officer. The data from the reports are then logged in the astronaut's health record and can be referred to for determining future flight eligibility.

Space Radiation Health Officer

One key member of the SRAG is the Space Radiation Health Officer (SRHO), whose job it is to interface with various directorates and research programs to provide the very best radiation assessment to the Chief Medical Officer (CMO), NASA Senior Management, and of course the crew. One of the SRHO's tasks in fulfilling this role is to keep up-to-date with research, especially research conducted within the realm of the Space Radiation Element of NASA's Human Research Program (HRP). The SRHO also interfaces with the Science Mission Directorate, which provides up-to-date information regarding the current radiation environment, the latest dosimetry measurements, and modeling of the radiation environment.

In support of the crew, the SRHO is responsible for inputting crew radiation data into the Lifetime Surveillance of Astronaut Health (LSAH)[2] record. The SRHO also draws upon outside expertise such as the National Council on Radiation Protection and Measurements (NCRP), the U. N. Scientific Committee on the Effects of

[2]The LSAH is a proactive occupational epidemiologically based health program for astronauts that screens and monitors crews for injury and/or disease related to their occupation. The program defines risks associated with the exposures encountered by astronauts, and using this data it refines follow-up medical examinations to capture sub-clinical medical events. The system helps detect potential health problems early and thereby help prevent progression to more serious disease.

Atomic Radiation (UNSCEAR), the International Commission on Radiological Protection (ICRP), the Radiation Effects Research Foundation (RERF), and the National Cancer Institute (NCR).

Radiation Risk

One of the SRAG team's tasks involves working with NASA to update the agency's Space Cancer Risk model (NSCR), which the team does by consulting data logged by the various radiation monitors [9]. The NSCR (Fig. 4.7) is a tool that:

1. Helps NASA estimate crew radiation risks during ISS missions.
2. Provides radiation data that can be used to estimate radiation risks during Exploration Class missions.
3. Helps NASA calculate the increase risk of cancer in the astronaut corps.
4. Provides NASA with up-to-date characterization of the GCR environment.
5. Provides age and gender specific radiation risks (Table 4.1).

For example, to consider gender effects, the SRAG team consults research (Table 4.2) that has investigated gender-specific differences and uses data from these studies to make an estimate of the relative risk for astronauts. For example, the SRAG team is able to use this data to make an estimate of the individual organ and tissue contributions to cancer risk for astronauts while on orbit (Table 4.3).

Fig. 4.7 Orion spacecraft. (Image courtesy of NASA)

Table 4.1 NASA radiation risk prediction model[a]

Epidemiological	Terrestrial research	Space radiation research
• BEIR—biological effects of ionizing radiation	• NIH/NCI—terrestrial cancer research	• Human research program
• UNSCEAR—U. N. scientific committee on the effects of atomic radiation	• DOE/DOD/DARPA—radiation effects research	• Space radiation focused research
• RERF—radiation effects research foundation	• International research activities	• Utilizes animal models with simulated space environment

[a]F. A. Cucinotta, L. Chappell, M. Y. Kim. Space radiation cancer risk projections and uncertainties – 2012. NASA/TP-2013-217375. 2013

Table 4.2 Sex-specific differences in the excess relative risk (ERR) per gray for major cancers[a]

Cancer type	ERR Gy^{-1} (averaged over both sexes)[b]	Female to Male Ratio (sex-specific ERR Gy^{-1} estimates)
All solid cancers	0.42	2.1[c]
Esophagus	0.60	4.3
Stomach	0.33	3.7
Colon	0.34	1.4
Liver	0.38	1.6
Gallbladder	0.48	0.43
Lung	0.75	2.7
Bladder	1.19	1.7

[a]Adapted from [10]
[b]The sex-averaged ERR Gy^{-1} is shown for subjects at the age of 70 year after exposure at 30 year
[c]A ratio of 2:1 can be interpreted as females having a risk of radiation induced cancer death that is 2.1 times that of males

Table 4.3 Individual organ and tissue contributions to cancer risk for astronauts at mid mission aged 47 years (solar minimum)

Males	Females
Lung (>20%)	*Lung* (>35%)
BFO (leukemia)	Stomach
Colon	BFO (leukemia)
Stomach	*Colon*
Bladder	*Ovarian*
Liver	*Breast*
Remainder Organs	Remainder organs
Prostate	Liver
Esophagus	Bladder
Brain	Brain
Oral mucosa	Esophagus
Skin	Uterus/cervix
Testes	Oral mucosa
Thyroid	Skin
	Thyroid

Protecting Crews During Solar Particle Events

The SRAG team is also involved in monitoring SPE activity and providing recommendations based on the SPE threat and potential impacts to mission management and the flight surgeon. Most of the information concerning SPE activity is obtained from a combination of data streamed from current assets, environmental assessment and solar state. In the event of an SPE, the radiation flight controller returns to his/her console in the mission control room and alerts the management and flight control team to ensure that the radiation monitoring system is available. If, after analysis of the data, the dose projection is considered negligible, no action is taken. But, if the SPE exceeds the NASA threshold, the crew is informed and is instructed to take cover in the more highly shielded areas of the ISS such as the service module, Node 2 crew quarters, and the U. S. lab.

Protecting Deep Space Crews

NASA's next manned vehicle will be the Orion, also known as the Multi-Purpose Crew Vehicle (MPCV). This vehicle will eventually ferry crews to destinations beyond Earth orbit and will therefore need to protect crews from the deep space environment. To obtain data on this environment the SRAG team performed radiation measurements during the Orion EFT-1 Mission, launched December 5, 2014. The EFT-1 trajectory included one low altitude orbit and one high altitude orbit that reached an altitude of almost 6000 km. While flying this profile, the Orion MPCV transited regions of intense proton and electrons, details of which were relayed to SRAG thanks to an independent radiation detector and a suite of optically and thermally stimulated luminescence detectors that were fitted into the vehicle.

References

1. Perez-Nunez D, Braby LA. Replacement tissue-equivalent proportional counter for the International Space Station. Radiat. Prot. Dosim. 2011;143(2-4):394–7.
2. Zhou D, Semones E, Weyland J. Radiation measured with TEPC and CR-39 PNTDs in low Earth orbit. Adv. Space Res. 2007;40(11):1571–4.
3. Narici L, Berger T, Matthia D, Reitz G. Radiation measurements performed with active detectors relevant for human space exploration. Front. Oncol. 2015;5:273.
4. Zhou D, Semones E, Gaza R, Zapp EN, Weyland J. Radiation measured for ISS-Expedition 12 with different dosimeters. Nucl. Instrum. Methods Phys. Res. Sect. A. 2007;580(3):1283–9.
5. Reitz G, Berger T, Bilski P, Facius R, Hajek M, Petrov VP, Puchalska M, Zhou D, Bossler J, Akatov YA, Shurshakov VA, Olko P, Ptaszliewicz M, Bergmann R, Fugger M, Vana N, Beaujean R, Burmeister S, Bartlett D, Hager L, Palfalvi JK, Szabó J, O'Sullivan D, Kitamura H, Uchihori Y, Yasuda N, Nagamatsu A, Tawara H, Benton ER, Gaza R, McKeever SW, Sawakuchi G, Yukihara EG, Cucinotta FA, Semones E, Zapp EN, Miller J, Dettmann

J. Astronaut's organ doses inferred from measurements in a human phantom outside the International Space Station. Radiat. Res. 2009;171(2):225–35.
6. Straube U, Berger T, Reitz G, Facius R, Fuglesang C, Reiter T, Damann V, Tognini M. Operational radiation protection for astronauts and cosmonauts and correlated activities of ESA medical operations. Acta Astronaut. 2010;66(7-8):963–73.
7. Spurny F, Jadrnickova I. Some recent measurements onboard spacecraft with passive detector. Radiat. Prot. Dosim. 2005;116(1-4):228–31.
8. Kodaira S, Kawashima H, Kitamura H, Kurano M, Uchihori Y, Yasuda N, Ogura K, Kobayashi I, Suzuki A, Koguchi Y, Akatov YA, Shurshakov VA, Tolochek RV, Krasheninnikova TK, Ukraintsev AD, Gureeva EA, Kuznetsov VN, Benton ER. Analysis of radiation dose variations measured by passive dosimeters onboard the International Space Station during the solar quiet period (2007-2008). Radiat. Meas. 2013;49:95–102.
9. Akopova AB, Manaseryan MM, Melkonyan AA, Tatikyan SS, Potapov Y. Radiation measurement on the International Space Station. Radiat. Meas. 2005;39:225–8.
10. Ozasa K, Shimizu Y, Suyama A, Kasagi F, Soda M, Grant EJ, Sakata R, Sugiyamaa H, Kodama K. Studies of the Mortality of Atomic Bomb Survivors, Report 14, 1950–2003: An Overview of Cancer and Noncancer Diseases Kotaro. Radiat. Res. 2012:177:229–43.

Chapter 5
Radiation Dosimetry and Detection

Radiation monitoring on board the International Space Station (ISS) is conducted to gather, analyze and characterize the radiation environment to better ensure crew health [1–3]. And thanks to careful consideration of radiation effects, the station does a very good job of protecting the crew. But given that astronauts are exposed to about 80 times the terrestrial radiation dose, radiation exposure still remains a limiting effect on an astronaut's career, which is why it is important to accurately monitor exposure as closely as possible.

Operations

To protect ISS crews from the effects of radiation, space agencies are guided by the ISS Medical Operations Requirement Document (MORD). The MORD identifies radiation exposure monitoring requirements as follows:

1. Radiation doses absorbed by human tissue
2. Charged particles and neutron radiation inside the ISS
3. Charged particles outside the ISS during spacewalks.

To quantify the station's internal and external radiation, NASA[1] has installed various *active* and *passive* radiation instruments to measure and document each astronauts' radiation exposure. In addition to the passive and active radiation instru-

[1] Career radiation exposures of astronauts are tracked by a NASA radiation specialist who logs radiation exposures for each astronaut in a document known as the Astronaut Annual Radiation Exposure Report.

The original version of this chapter was revised. The correction to this chapter is available at https://doi.org/10.1007/978-3-319-74615-9_11

E. Seedhouse, *Space Radiation and Astronaut Safety*, SpringerBriefs in Space Development, https://doi.org/10.1007/978-3-319-74615-9_5

ments, each astronaut is provided with a dosimeter that serves as the dosimeter of record [4, 5]. The data from each astronauts' dosimeter, when combined with data from internal and external dosimeters, provides an accurate characterization of the radiation environment

Exposure Limits

Terrestrial exposure limits (see Chap. 3) are much too restrictive if applied to the space environment. So space agencies have adopted recommendations made by the National Council on Radiation Protection (NCRP) that sets a career limit at a three percent Risk of Exposure-Induced Death (REID) from cancer. But how does this number compare with the incidence of cancer in the general population? Well, here on Earth, the risk of developing and dying from cancer is about 20%, which means about 20 people out of 100 will die from cancer. But if you happen to be an astronaut then that risk increases by three percent which means 23 astronauts out of every 100 astronauts may die [6].

Passive Dosimetry

One way of measuring and tracking radiation is to use passive dosimeters. These are placed at fixed locations around the ISS inside the pressurized modules. The information from these dosimeters provides the ground with information about those locations where the exposure rate is high. This information can then be used to re-evaluate the amount of time each astronaut spends in each module [3].

Passive Radiation Dosimeter

Passive Radiation Dosimeters (PRDs) are also known as Radiation Area Monitors (RAMs). Each RAM (Figs. 5.1 and 5.2) comprises a set of thermoluminescent detectors (TLDs) surrounded by Lexan. The material reacts to radiation by means of excitation of the materials that make up the TLDs. TLDs comprise lithium fluoride or calcium fluoride embedded in a solid crystal structure. When the TLD is exposed to radiation, the radiation interacts with the crystal. Some of the atoms in the crystal absorb the energy and become ionized. This generates free electrons and heating of the crystal which causes the material to vibrate which in turn releases the electrons. When the electrons return to their original pre-ionized state they release the stored energy which appears as light. This light is measured using special photomultiplier tubes and the amount of light—the 'glow curve'—released is proportional to the amount of radiation that struck the crystal [3, 4, 7].

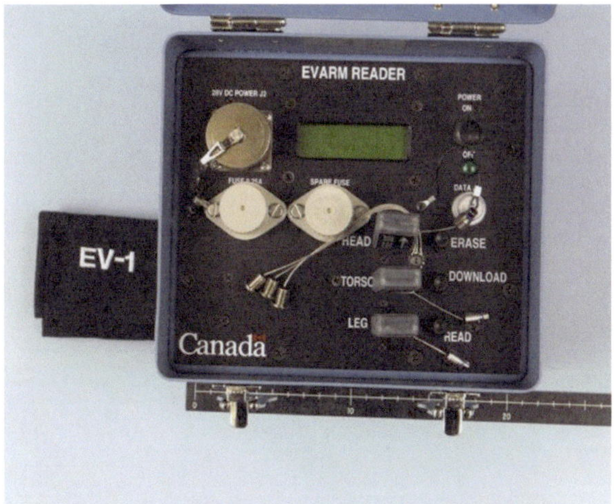

Fig. 5.1 A Canadian Space Agency radiation dosimeter used during spacewalks. (Image courtesy of CSA)

ISS002E6593 2001/06/07 12:48:13

Fig. 5.2 View of radiation area monitor (RAM), dosimeter. (Image courtesy of NASA)

Crew Passive Dosimeter

Crew Passive Dosimeters (CPDs) are provided to each astronaut. Apart from sporting a different label, the CPDs (Fig. 5.3) are exactly the same as the RAMs. They are worn by the astronauts throughout the mission, including spacewalks.

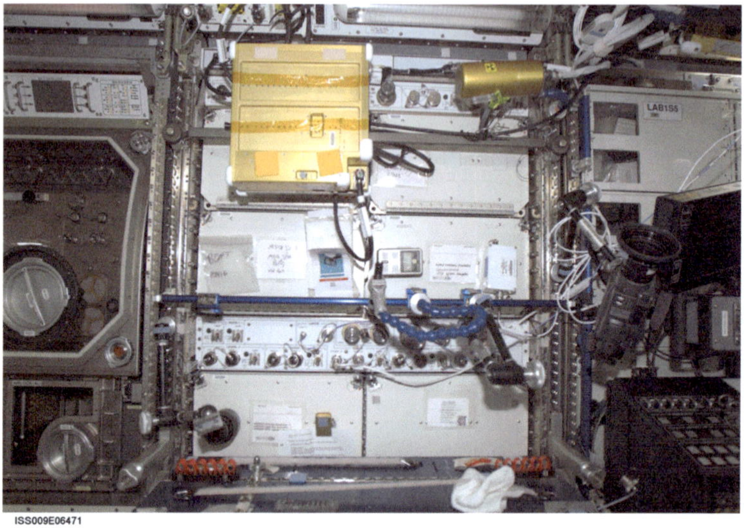

Fig. 5.3 EXPRESS Rack 5 with the intravehicular charged particle directional spectrometer (IV-CPDS) (gold box in left field of view), Tissue equivalent proportional counter (TEPC) radiation detector (gold cylinder) and upper storage compartments visible. EXPRESS Rack 5 is in the Destiny/U.S. laboratory. Image taken during Expedition 9. (Image courtesy of NASA)

Active Dosimetry

Active radiation monitors provide data to the ground that is used together with the data provided by the CPDs to generate high dose rate and low dose rate areas inside the station [4, 8, 9]. These measurements also help the ground to reduce uncertainty when calculating risk assessments for the crew, which is done by evaluating the following metrics:

1. Linear Energy Transfer spectra inside the ISS
2. Mass distribution
3. Space weather conditions
4. Stage of the solar cycle
5. CPD data
6. RAM data

Tissue Equivalent Proportional Counter

The Tissue Equivalent Proportional Counter (TEPC) uses gas to measure the dose of radiation. The function of the TEPC (Fig. 5.4) is to develop an exposure history of the crew during their stay on station. It does this by collecting data by making spectral measurements of the energy loss of radiation as the radiation passes through the detector volume. Inside the TEPC is an omni-directional detector encased in

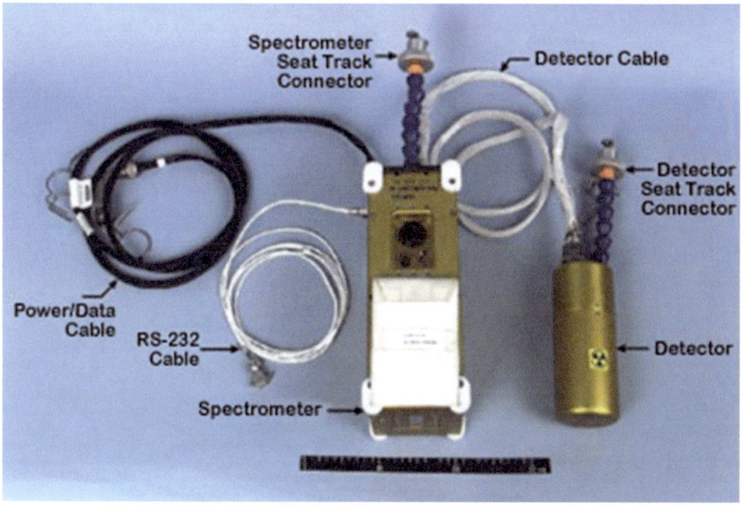

Fig. 5.4 Tissue equivalent proportional counter. (Image courtesy of NASA)

tissue-equivalent plastic similar to that used in the Matroshka Human Phantom [10], discussed later in this chapter. Also inside the TEPC is propane gas that, combined with the plastic, provides an energy deposition effect similar to human tissue. The way this energy deposition response is achieved is thanks to the propane gas which is kept at very low pressure. This means that the radiation passing through the detector gas passes through with similar linear energy loss as a human cell. This information is then stored inside a 512-channel spectrometer that is presented to the crew via an electronic display that displays total dose, total dose equivalent and incremental dose. The TEPC also downlinks information to the ground for analysis. The TEPC became operational in 2000 and as of 2017 it is still functioning (there is one active and one spare unit on board the ISS).

Charged Particle Directional Spectrometer

The Charged Particle Directional Spectrometer (CPDS) monitors the internal radiation environment of the ISS using a Cerenkov Detector. The Cerenkov Detector works by measuring Cerenkov light. The principle by which the detector works is as follows. A particle that passes through material (in this case a 12-element silicon stack) at a speed faster than the speed at which light can pass through the material emits light. An analogy is the sonic boom generated by an aircraft moving through the air faster than the sound waves can move through the air. In the case of the Cerenkov Detector, the amount of radiation can be calculated if the angle *and* direction of light is known. Three CPDS systems have been deployed externally: EV1-CPDS and EV3-CPDS were aligned with the +X and −X axis while EV2-CPDS was aligned with the −Z axis.

Intra-Vehicular TEPC

The Intra-vehicular TEPC (IV-TEPC) device measures radiation in near real time [8]. If the level of radiation exceeds pre-determined thresholds, the device signals a Caution and Warning (C&W) alarm. In such an event the crew would probably relocate to a higher shielded module. The unit itself is portable and comprises several tissue equivalent radiation detectors constructed of material and gas that react to radiation in a similar way that human tissue does. The IV-TEPC provides the following capabilities [8]:

1. Signal conditioning
2. Data manipulation
3. Storage
4. Real-time telemetry
5. Extended data download

The IV-TEPC was declared operational on board the ISS in March 2001. It failed in 2006 and has not been replaced.

European Crew Personal Active Dosimeter

The European Crew Personal Active Dosimeter (EuCPAD) is a device that assesses radiation exposure of European astronauts (Figs. 5.5 and 5.6). The device comprises three main elements:

1. Mobile Unit
   Two silicon detector modules
   Absorbed dose detector
2. Personal Storage Device
   TEPC
   Internal Mobile Unit
   Storage and charging capability for Mobile Unit
   Local data analysis
   Display of radiation data

Fig. 5.5 Wearable units of the new European crew personal active dosimeter—EuCPAD—system for active radiation monitoring of astronauts in orbit. (Image courtesy of ESA)

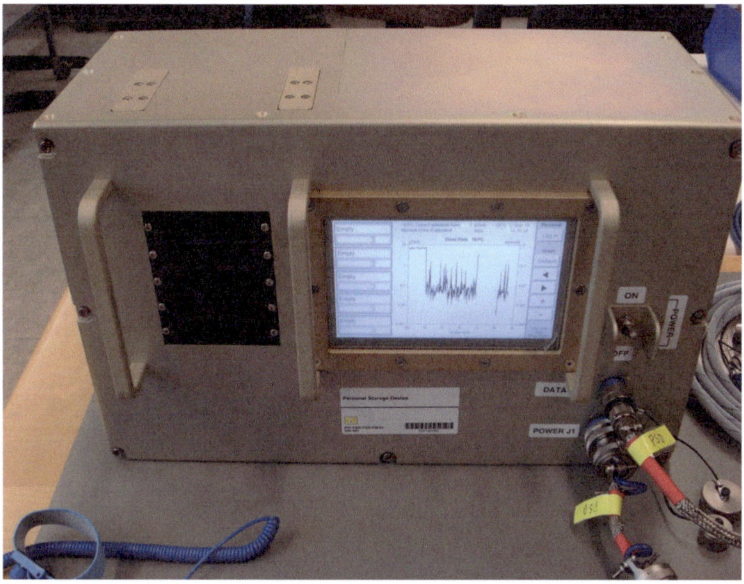

Fig. 5.6 EuCPAD is comprised of wearable units plus a personal storage device (PSD) for battery charging and data transfer, similar in purpose to a smartphone charging station. (Image courtesy of ESA)

3. Ground Station Analysis Software
   Calculates the dose equivalent based on spectroscopic data.

The Mobile Units (MUs) are battery driven and have a power and data interface with Columbus, where astronauts can access radiation data and where European Crew Active Dosimetry Activity is monitored in conjunction with the ground.

PADLES

The Passive Dosimeter for Life-Science Experiments in Space (PADLES) is used to monitor radiation inside the Japanese Experiment Module (*Kibo*) (Fig. 5.7). *Bio* PADLES is used to measure the biological effects of radiation while *Area* PADLES is used to assess the personal exposure of each astronaut, and *Free-Space* PADLES measures the radiation environment outside Kibo. The key radiation sensitive material inside the detectors is CR-39 plastic and the radiation measured by the dosimeters is published in JAXA's PADLES database.[2] Operationally, *Free-Space* PADLES is launched and returned via pressurized cargo on board the Cygnus or Dragon, packed inside a Crew Transfer Bag. On station, *Free-Space* PADLES is

[2] http://idb.exst.jaxa.jp/db_data/padles/NI005.html.

Fig. 5.7 Japanese astronaut Satoshi Furukawa holds the PADLES detector kit in the Unity module. (Image courtesy of NASA)

installed on the Multi-Purpose Experiment Platform inside the JEM. It is then moved outside the JEM via the JEM airlock and is grabbed by the JEM RMS robot arm which positions *Free-Space* PADLES on the exterior of the module.

ISS Detectors

The ISS flies at an altitude of around 400 km. At this altitude, crews are provided with a safe space platform that is protected against most of the deleterious effects of ionizing radiation. Over the years of ISS operation, the best *active* dosimeters have proven to be those that provide Linear Energy Transfer (LET) data and tissue equivalent proportional data (such as the TEPCs). Supporting the data provided by the dosimeters is a suite of detectors. Together, these dosimeters and detectors are used to provide a very accurate set of baseline data that allow scientists to benchmark risk assessments for long duration flights [9, 11]. As long as those flights occur in LEO. Beyond LEO? Well, that's another story.

R-16

This was one of the first radiation detectors to be used on board the ISS.

DB-8

This detector, which features two independent sensors, was located in the Zvezda module in the early stages of ISS construction.

Liulin

The Liulin system utilizes silicon detectors to measure deposited radiation energy [12–16]. Basically, the number of charged particles that hit the device converts to a dose rate. The first version (Liulin-E094) was first used on board the ISS in April 2001 and was followed by a series of updated systems that were flown inside the ISS (Liulin ISS between September 2005 and June 2014 and Liulin-5, which was deployed between May 2007 to present) and outside the ISS (R3DE from February 2008 to September 2009 and R3DR between March 2009 and August 2010).

Alteino

This detector is also referred to as SilEye3. It made its first appearance on the *Mir* space station. The shoebox-sized system is comprised of a stack of eight silicon striped sensors measuring 80 × 80 × 0.38 mm and two plastic scintillators [17–19]. The orientation of the stack is configured to provide a set of three coordinates through which particles strike. This configuration allows for tracking the direction of particles.

ALTEA

This is a system of six silicon telescopes similar to Alteino except for the scintillators. These detectors have been positioned in the U. S. lab in three axes and also in the Columbus module [20, 21]. The system downlinks data via real time telemetry.

DosTel

This detector system is also based on silicon detectors. They are deployed in the Columbus module in the X and Y axes [22, 23].

TRITEL

This system is comprised of a set of three silicon telescopes configured in a three-dimensional arrangement. The system has been deployed in the Columbus module (in the European Physiology Module rack TRITEL-SURE) on ISS since 2012.

PAMELA

PAMELA is designed to study antimatter [24, 25]. To do this, scientists need information about the particle charge, energy, and interaction of those particles with other particles. The system is comprised of a magnetic spectrometer, a sampling electromagnetic calorimeter made up of 44 silicon sensor layers, and a neutron detector.

RAD

The RAD is not mounted anywhere on the ISS, and it will not find a place on any future manned spacecraft. It is mentioned here because the data generated by the RAD will be used to plan future manned missions beyond low Earth orbit. The detector was mounted on *Curiosity* during its transit to Mars, during which time it sent back data about the deep space radiation environment [26–29]. After landing, the system continued to provide detailed measurements of the surface radiation environment.

Matroshka

Measuring radiation hazards in space—MATROSHKA—is an experiment sponsored by Roscosmos, ESA, and JAXA. It is comprised of a human phantom that measures the radiation astronauts are exposed to inside and outside the ISS [30–32]. The phantom is designed using human tissue-equivalent material and is filled with water and instrumented with a series of passive radiation detectors (Fig. 5.8).

The phantom is essentially a torso comprising 33 slices each of 2.5-cm thickness. Each slice houses thermoluminscent lithium fluoride detectors (about 4.5 mm in diameter) placed in plastic tubes. Thanks to the positioning of the TLDs inside the phantom it is possible for scientists to accurately determine the spatial distribution of radiation and thereby calculate the effective dose.

The key to the way radiation is measured is found inside the TLDs. Inside each detector is a lattice that traps free electrons created by radiation. The greater the radiation dose the greater the number of trapped electrons [10]. When exposed to

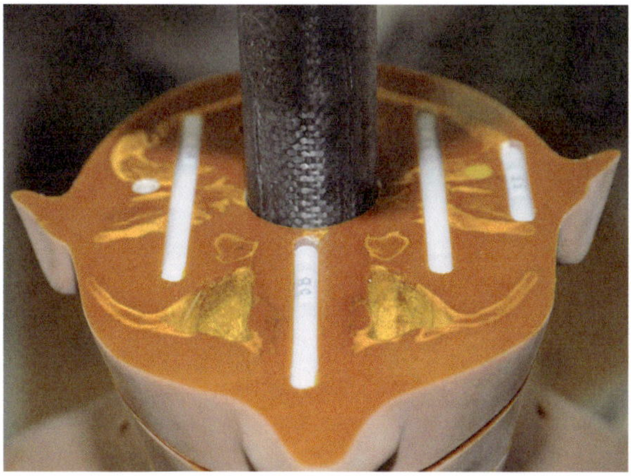

Fig. 5.8 The outcome of the MATROSHKA experiment was the first comprehensive measurement of long-term exposure of astronauts to cosmic radiation. The experiment was conducted on board and outside the ISS. To measure the radiation doses that astronauts are exposed to, ESA in collaboration with research institutions from Germany, Poland, Austria, Sweden and Russia, designed and carried out the MATROSHKA experiment. MATROSHKA is a phantom that closely resembles the human body. Made of tissue equivalent material, the phantom is packed with thousands of detectors that recorded the doses from cosmic radiation over several years. (Image courtesy of NASA)

heat, the trapped electrons are released, emitting light, and it is this light that provides the index for radiation exposure: the greater the light the higher the proportional radiation dose. Between 2004 and 2009 MATROSHKA was exposed to radiation on three occasions: two inside the Russian orbital segment and one outside the ISS. After crunching the numbers, scientists at the German Space Agency and the Technical University in Vienna revealed that individual dosimeters worn by astronauts inside ISS had overestimated radiation exposure by 15% compared with the actual dose measured inside MATROSHKA. The overestimation in space was more than 200%.

> *We must remember that measurements within the MATROSHKA experiment were performed at low Earth orbit where the Earth's magnetosphere significantly reduces the number of charged particles from cosmic radiation. In interplanetary space there is no such shielding.*

Dr. Bilski, a scientist who suggested that manned Mars missions may still be a risky proposition despite the lower than expected radiation levels measured on ISS.

References

1. Koontz SL, Boeder PA, Pankop C, Reddell B. The ionizing radiation environment on the International Space Station: performance vs. expectations for avionics and materials. 2005 IEEE Radiation Effects Data Workshop; 2005 Jul 11–15. pp. 110–116.
2. Akopova AB, Manaseryan MM, Melkonyan AA, Tatikyan SS, Potapov Y. Radiation measurement on the International Space Station. Radiat Meas. 2005;39:225–8.

3. Kodaira S, Kawashima H, Kitamura H, Kurano M, Uchihori Y, Yasuda N, Ogura K, Kobayashi I, Suzuki A, Koguchi Y, Akatov YA, Shurshakov VA, Tolochek RV, Krasheninnikova TK, Ukraintsev AD, Gureeva EA, Kuznetsov VN, Benton ER. Analysis of radiation dose variations measured by passive dosimeters onboard the International Space Station during the solar quiet period (2007–2008). Radiat Meas. 2013;49:95–102.
4. Zhou D, Semones E, Gaza R, Johnson S, Zapp EN, Weyland M. Radiation measured for ISS-Expedition 12 with different dosimeters. Nucl Instrum Methods Phys Res Sect A. 2007;580(3):1283–9.
5. Deme S, Apathy I, Pazmandi T, Benton ER, Reitz G, Akatov YA. On-board TLD measurements on MIR and ISS. Radiat Prot Dosim. 2006;120(1-4):438–41.
6. Narici L, Berger T, Matthia D, Reitz G. Radiation measurements performed with active detectors relevant for human space exploration. Front Oncol. 2015;5(273):10.
7. Dachev TP, Atwell W, Semones E, Tomov BT, Reddell B. Observations of the SAA radiation distribution by liulin-E094 instrument on ISS. Adv Space Res. 2006;37(9):1672–7.
8. Zhou D, Semones E, Weyland M, Johnson S. Radiation measured with TEPC and CR-39 PNTDs in low Earth orbit. Adv Space Res. 2007;40(11):1571–4.
9. Lishnevskii AE, Panasyuk MI, Benghin VV, Petrov VM, Volkov AN, Nechaev OY. Variations of radiation environment on the International Space Station in 2005-2009. Cosm Res. 2012;50(4):319–23.
10. Reitz G, Berger T, Bilski P, Facius R, Hajek M, Petrov VP, Puchalska M, Zhou D, Bossler J, Akatov YA, Shurshakov VA, Olko P, Ptaszliewicz M, Bergmann R, Fugger M, Vana N, Beaujean R, Burmeister S, Bartlett D, Hager L, Palfalvi JK, Szabó J, O'Sullivan D, Kitamura H, Uchihori Y, Yasuda N, Nagamatsu A, Tawara H, Benton ER, Gaza R, McKeever SW, Sawakuchi G, Yukihara EG, Cucinotta FA, Semones E, Zapp EN, Miller J, Dettmann J. Astronaut's organ doses inferred from measurements in a human phantom outside the International Space Station. Radiat Res. 2009;171(2):225–35.
11. Lishnevskii A, Panasyuk M, Benghin V, Petrov VM, Volkov AN, Nechayev OY. Variations of radiation environment onboard the ISS in the year 2008. Cosm Res. 2010;48:206–10.
12. Dachev T, Tomov B, Matviichuk Y, Dimitrov P, Lemaire J, Gregoire G. Calibration results obtained with Liulin-4 type dosimeters. Adv Space Res. 2002;30:917–25.
13. Dachev TP, Tomov BT, Matviichuk YN, Dimitrov PG, Spurny F, Ploc O. Liulin type spectrometry-dosimetry instruments. Radiat Prot Dosim. 2011;144:675–9.
14. Dachev TP, Spurny F, Ploc O. Characterization of radiation environment by Liulin type spectrometers. Radiat Prot Dosim. 2011;144:680–3.
15. Dachev TP, Semkova JV, Tomov BT, Matviichuk YN, Maltchev PGS, Koleva R, et al. Overview of the Liulin type instruments for space radiation measurement and their scientific results. Life Sci Space Res. 2015;4:92–114.
16. Semkova J, Koleva R, Shurshakov V, Benghin V, Maltchev S, Kanchev N, et al. Status and calibration results of Liulin-5 charged particle telescope designed for radiation measurements in a human phantom onboard the ISS. Adv Space Res. 2007;40:1586–92.
17. Casolino M, Bidoli V, Furano G, Minori M, Morselli A, Narici L, et al. The Sileye-3/Alteino experiment on board the International Space Station. Nucl Phys B Proc Suppl. 2002;113:71–8.
18. Larsson O, Benghin VV, Berger T, Casolino M, Di Fino L, Fuglesang C, et al. Measurements of heavy-ion anisotropy and dose rates in the Russian section of the International Space Station with the Sileye3/Alteino detector. J Phys G Nucl Part Phys. 2015;42:025002.
19. Narici L, Belli F, Bidoli V, Casolino M, De Pascale MP, Di Fino L, et al. The ALTEA/Alteino projects: studying functional effects of microgravity and cosmic radiation. Adv Space Res. 2004;33:1352–7.
20. Narici L, Bidoli V, Casolino M, De Pascale MP, Furano G, Morselli A, et al. ALTEA: anomalous long-term effects in astronauts. A probe on the influence of cosmic radiation and microgravity on the central nervous system during long flights. Adv Space Res. 2003;31:141–6.
21. Di Fino L, Casolino M, De Santis C, Larosa M, La Tessa C, Narici L, et al. Heavy ions anisotropy measured by ALTEA in the International Space Station. Radiat Res. 2011;176:397–406.

22. Beaujean R, Kopp J, Burmeister S, Petersen F, Reitz G. Dosimetry inside Mir station using a silicon detector telescope (DOSTEL). Radiat Meas. 2002;35:433–8.
23. Labrenz J, Burmeister S, Berger T, Heber B, Reitz G. Matroshka DOSTEL measurements onboard the International Space Station (ISS). J Space Weather Space Clim. 2015;5:A38.
24. Adriani O, Barbarino GC, Bazilevskaya GA, Bellotti R, Boezio M, Bogomolov EA, et al. The PAMELA mission: heralding a new era in precision cosmic ray physics. Phys Rep. 2014;544:323–70.
25. Adriani O, Barbarino GC, Bazilevskaya GA, Bellotti R, Boezio M, Bogomolov EA, et al. A statistical procedure for the identification of positrons in the PAMELA experiment. Astropart Phys. 2010;34:1–11.
26. Hassler DM, Zeitlin C, Wimmer-Schweingruber RF, Böttcher S, Martin C, Andrews J, et al. The radiation assessment detector (RAD) investigation. Space Sci Rev. 2012;170:503–58.
27. Zeitlin C, Hassler DM, Cucinotta FA, Ehresmann B, Wimmer-Schweingruber RF, Brinza DE, et al. Measurements of energetic particle radiation in transit to Mars on the Mars science laboratory. Science. 2013;340:1080–4.
28. Hassler DM, Zeitlin C, Wimmer-Schweingruber RF, Ehresmann B, Rafkin S, Eigenbrode JL, et al. Mars' surface radiation environment measured with the Mars science laboratory's curiosity rover. Science. 2014;343:1244797.
29. Guo J, Zeitlin C, Wimmer-Schweingruber RF, Hassler DM, Posner A, Heber B, et al. Variations of dose rate observed by MSL/RAD in transit to Mars. Astron Astrophys. 2015;577:A58.
30. Semkova J, Koleva R. Overview on radiation quantities observed by Liulin-5 instrument in a human phantom on the International Space Station during the minimum of the 23rd solar cycle. C R Acad Bulg Sci. 2010;63:1533–42.
31. Semkova J, Koleva R, Maltchev S, Benghin V, Shurshakov V, Chernykh I, et al. Preliminary results of Liulin-5 experiment for investigation of the dynamics of radiation doses distribution in a human phantom aboard the International Space Station. C R Acad Bulg Sci. 2008;61:787–94.
32. Reitz G, Berger T, The MATROSHKA Facility—Dose determination during an EVA. Radiat. Prot. Dosimetry. 2006;120:442–5.

Chapter 6
Shielding

The space radiation environment will be a critical consideration for everything in the astronauts' daily lives, both on the journeys between Earth and Mars and on the surface. You're constantly being bombarded by some amount of radiation.

– Ruthan Lewis, architect and engineer at NASA's Goddard Space Flight Center

August 7, 1972. An enormous flare explodes from the Sun, spewing out a burst of energetic particles. A Moonwalker caught in the storm would have been exposed to 400 rem. Not necessarily deadly, but enough to require a mission abort and an early return to Earth. Fortunately there were no astronauts on the surface of the Moon in August 1972. *Apollo 16* had returned to Earth the previous month, and the crew of Apollo 17 was preparing for their mission that was due to take place in December that year [1]. Of course, an astronaut isn't going to be wandering around on the Moon when a storm hits. They will be ensconced inside their base or spacecraft. If such an event had occurred during an Apollo mission the Apollo command module's hull[1] would have reduced the 400 rem to about 35–40 rem. Still enough of a dose to cause a headache and perhaps some nausea, but not sufficient to require a bone marrow transplant. Jonathan Pellish, space radiation engineer at Goddard, says: "There's a lot of good science to be done on Mars, but a trip to interplanetary space carries more radiation risk than working in low-Earth orbit."

In science fiction movies, the most dangerous threat to the crew is usually some form of alien life. And in most such movies these threats are usually pretty big. But in the real world of sending astronauts on deep space interplanetary missions (Figs. 6.1 and 6.2) the dangers are mostly invisible. Heavy elementary particles zipping along through space and tearing through DNA is enough to give any astronaut serious concern [2, 3]. These cosmic rays present irreducible risks and are as deadly as any threat Hollywood can conjure up.

[1] The Apollo Command Module's hull provided 8 g/cm^2 of radiation protection. The space shuttle had 11 g/cm^2, and the International Space Station has up to 15 g/cm^2 in its most shielded areas. In contrast, a spacesuit has only 0.25 g/cm^2.

E. Seedhouse, *Space Radiation and Astronaut Safety*, SpringerBriefs in Space Development, https://doi.org/10.1007/978-3-319-74615-9_6

Fig. 6.1 Shielding will be critical for those venturing beyond Earth orbit. (Image courtesy of NASA)

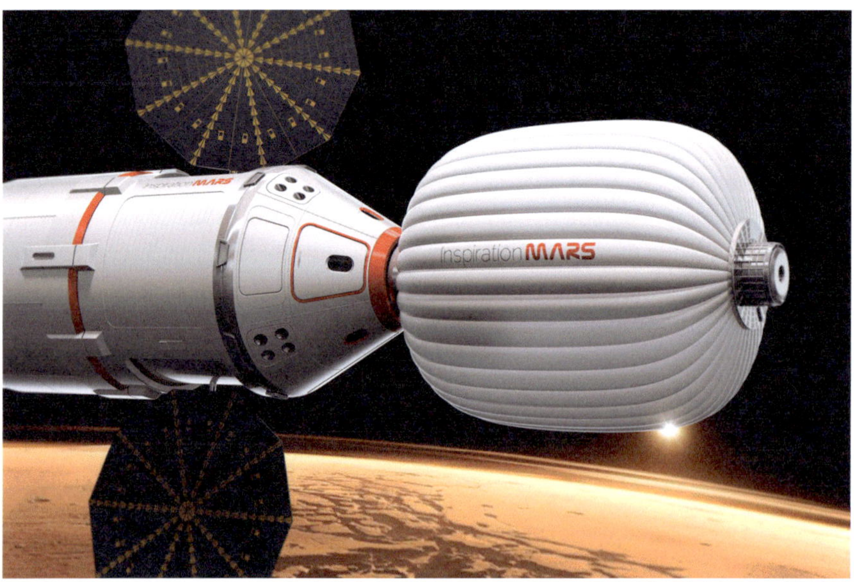

Fig. 6.2 Astronauts venturing to Mars will need shielding—a lot of it! (Image courtesy of Credit Inspiration Mars)

Here on Earth, the sheer bulk of the atmosphere does a great job shielding us from the worst those shrapnel-like cosmic rays can inflict. Many miles above us, incoming protons are absorbed by the nuclei of air atoms. Particles and sub-particles disperse in a series of annihilating cycles until all that is left are some pions and mesons, some of which pass through our body. But at that stage, there is so little energy left in them—thanks to the weight of the atmosphere—that all they can do is produce a few ions.

Above the atmosphere and beyond low Earth orbit, the situation is different. In deep space, those same cosmic rays have nothing to disperse them, except spacecraft and the astronauts inside them. And once those heavy nuclei zip through the skin of the spacecraft and through the human body, the trail of damage is devastating—broken bonds, genetic material ripped apart, tissues permanently damaged [4, 5]. The body has an extraordinary capacity for self-repair, so a week or month of this is survivable. But two or more years? No way. And we know this from studying the grave biological consequences of the unfortunate humans who have been exposed to intense bursts of radiation.

The Mars evangelists promote the fact that since some astronauts have spent six months in space, traveling to Mars should be a breeze. But astronauts on board the International Space Station (ISS) are shielded by Earth's magnetic field. Solutions? Shielding perhaps? One shield suggested by those in the business of protecting astronauts during exploration class missions (ECMs) is a sphere of water. The only drawback is that such a system would weigh 400 tons. A superconducting magnet perhaps? Such a system (Fig. 6.3) would use a magnetic field to repel cosmic rays, but the problem is that the magnetic field would present certain health risks.

So what other shielding solutions are out there? Well, before we discuss these, it's important to have a reference for what exactly engineers are up against. We'll begin with the legal limit for those working in nuclear power plants. That number happens to be 5 rems per year. A Mars astronaut by comparison would be exposed to 80 rems per year, and the consequences of that exposure would be that one in ten male interplanetary astronauts would die from cancer. Many more would suffer from radiation-induced cataracts and brain damage [6]. And that's a best case scenario, because it isn't just cosmic rays that inflict damage. Every once in a while, the Sun unleashes huge bursts of heavy nuclei that travel at close to light speed. These bursts, which can deliver more than a 100 rem/h, are, quite simply, a death sentence for any unshielded astronaut in deep space.

Water as a Radiation Shield

So what about those options mentioned earlier? Before considering the use of water (Figs. 6.4 and 6.5) (astronauts need this, so it makes sense to use it as a shielding material) as a shielding material we need to perform some basic calculations. First, we need to know how much shielding material it takes to protect an astronaut.

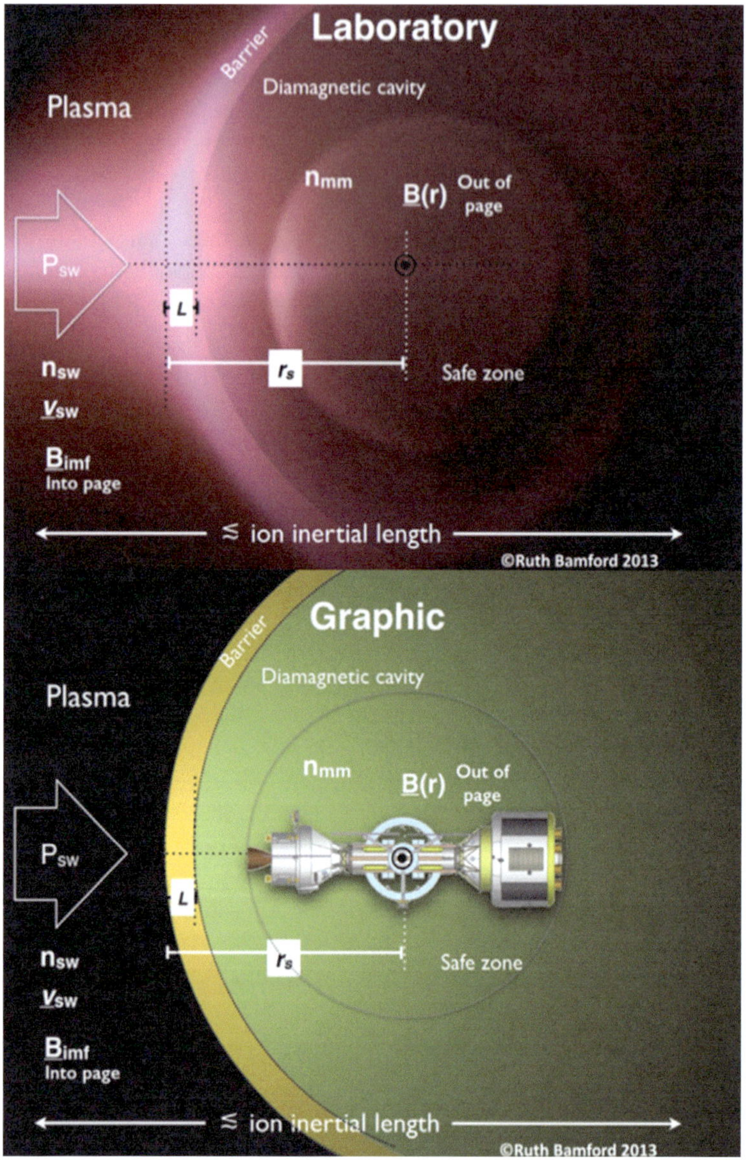

Fig. 6.3 NASA has a 158-page report on superconducting radiation shielding. Magnetic fields with 10 T and 10 m thick could deflect about 93% of the radiation—in theory. (Image courtesy of NASA)

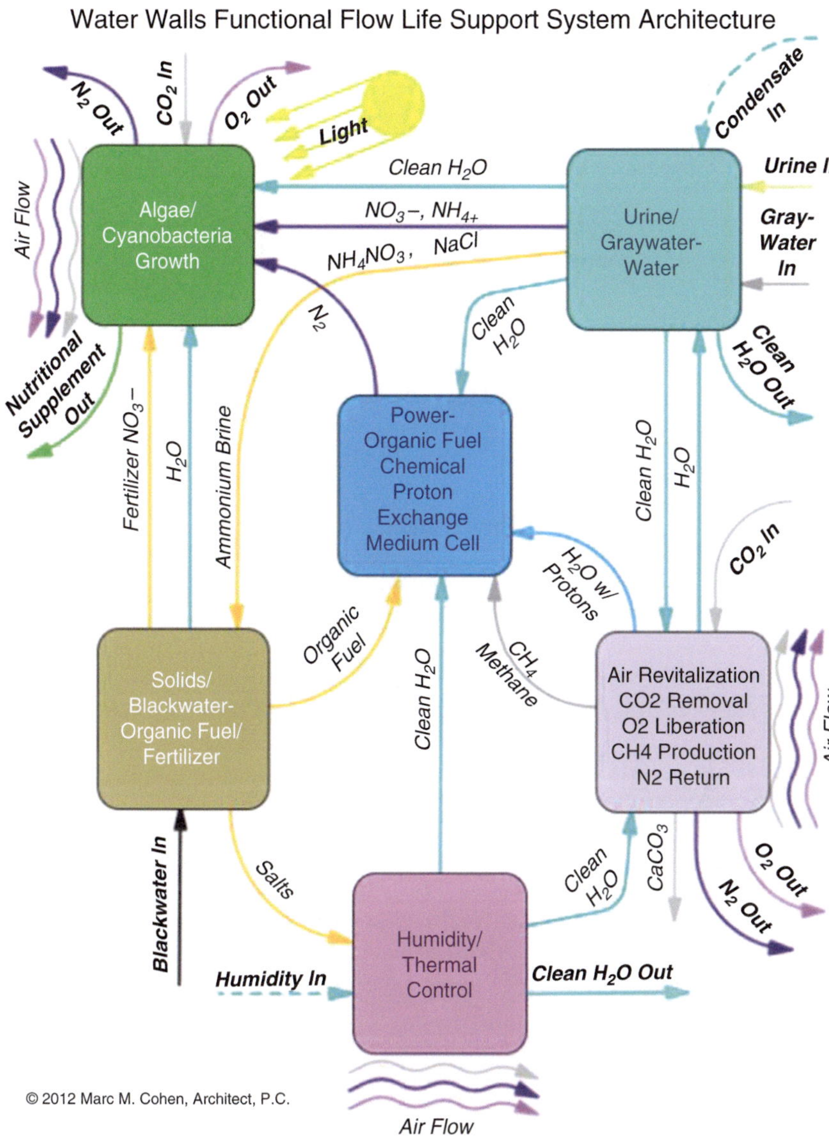

Fig. 6.4 Water wall system architecture. (Image courtesy of Marc Cohen. Used with permission)

If we wanted to provide an interplanetary astronaut with the same shielding as on Earth it would take one kilogram of water per square centimeter. But, since astronauts are willing to accept risk, let's give them less—and more affordable—protection and have them make do with just 500 g of water per square centimeter. That amount of shielding is equivalent to you living at an altitude of 5500 m. Now for the sake of simplicity, let's make our spacecraft a sphere. To protect our crew with water, the walls of

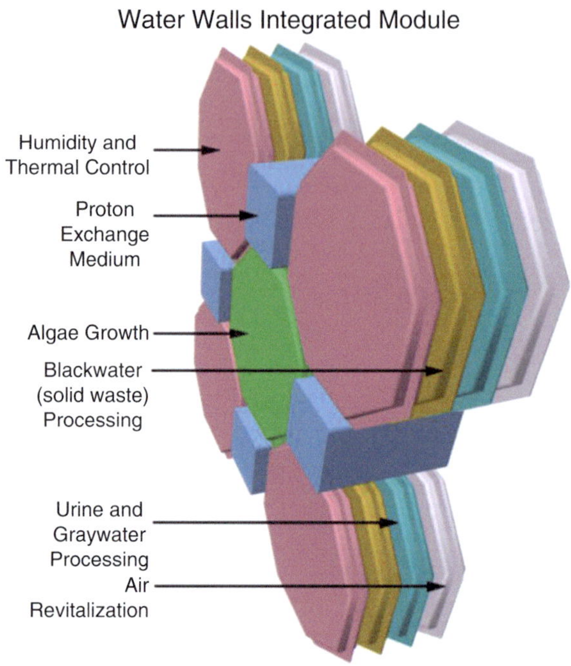

Fig. 6.5 Water walls integrated module. (Image courtesy of Marc Cohen. Used with permission)

this spherical vehicle would need to be 5 m thick and would weigh about 500 tons. Back in the old days, the space shuttle could ferry around 30 tons into LEO. The Space Launch System? About 130 tons in its most powerful lift configuration.

So 500 tons is too heavy. But what if the engineers reduced the amount of water and increased the *hydrogen* content of the spacecraft walls (Fig. 6.5)? They could do this by using polyethylene and perhaps bring the weight down to 400 tons. That just isn't financially feasible, so let's consider the other option mentioned—magnetic shielding.

Magnetic Shielding

This is yet another exotic shielding option promoted by the Mars evangelists. 'Shielding?' they say. Let's use magnetic shielding [7]. Problem solved! Well, no, the problem is not solved because this method of shielding hasn't moved far beyond the PowerPoint phase of development, and there are good reasons for this. Earth is surrounded by a magnetic field that does a great job deflecting incoming charged particles, so it would seem reasonable to assume—unless you happen to be a particle physicist—a spacecraft could carry a magnet to do the same. The problem involves those cosmic rays. They have tremendous kinetic energy, and trying to

Fig. 6.6 An artistic representation of an active, magnetic, toroidal shield used for protecting astronauts from astroparticles during deep space missions. (Image courtesy of Giorgina Colleoni & Valerio Calvelli European Commission)

bring them to a standstill in the space of just a couple of meters requires energy the likes of which might be achievable in the world of *Star Trek* but in the real world is not. That's because it would require a magnetic field of 20 T to stop cosmic rays, and 20 T is about 600,000 times the strength of Earth's magnetic field. How would humans endure living in a magnetic field of 20 T and what would the long-term effects be? Volunteers form a line here please! But, the Mars evangelists insist, it might be possible to use a second magnet to cancel out the field effect of the first magnet. Such a system, its proponents argue, could use plasma to push away the magnetic field of the first. But the problem with plasma is that is very unstable, and even if it could be controlled, the nuances of how plasma behaves in a magnetic field (Fig. 6.6) would mean the field would be weakened, not strengthened.

Electrical Shielding

So, water is too expensive and the magnetic option is too tricky—and downright dangerous. What about electrical fields? In this application, the spacecraft would be charged electrically…with *2 billion volts!* Such a charge would, in theory, repel cosmic ray protons. The problem is that space—even deep space—is not empty.

Even in deep space there are ions and electrons flying around, and these negatively charged electrons would be attracted by a spacecraft that is positively charged [8]. Let's not forget that this spacecraft would have an electric field that would extend tens of thousands of kilometers away from the vehicle. Such a huge electric field would draw in electrons from a huge volume of space, and when those electrons hit the walls of the spacecraft they would act just like the cosmic rays the shield was designed to repel! The electrons would generate gamma rays as soon as they hit the vehicle, and the intensity of this barrage would be so great that it would put the original headache in the shade. And what about those 2 billion volts? Does anyone have any idea of what sort of system could generate such a current? Two billion volts is 2000 MW, which is about the same amount of power generated by your average power plant. How do you fit such a system on an interplanetary spacecraft? The answers to these questions are few and far between, but it doesn't seem to have discouraged the grant providers who seem to continue to throw money at the idea.

Linear Energy Transfer and Relative Biological Effectiveness

One process that is key to understanding which materials make the best shields is how radiation interacts with the spacecraft and the occupants inside. That is because radiation does not simply pass through the walls of a spacecraft, just like it does not simply pass through the bodies of the astronauts. Radiation *interacts*. And as it interacts (see boxed texts later), the energy of all that ionizing radiation is disrupted, and the size of the particles reduced [9–12]. Which is a good thing. The problem is that the disruption causes the heavy charged particles—the *primary radiation*—to disintegrate into smaller particles—*secondary radiation*—and it is these smaller particles that cause biological damage in the crew.

Now you might think the solution would be to conduct research that mimics this interactive process, but that is not what research does. Instead, almost all research studies that simulate the effect of galactic cosmic rays (GCRs) do so by exposing animals to heavy-ion accelerators to simply replicate the dose a human crew might be exposed to during an interplanetary mission. But this method does not provide a true model of what happens in deep space because it is very difficult to replicate the myriad energies of disrupted GCRs and even more difficult to measure the extent to which these energies inhibit cell regrowth and tissue repair mechanisms. Furthermore, different animals respond differently to radiation. Some are more susceptible and some less sensitive to radiation damage. And finally, the current technology of heavy-ion accelerators is limiting in terms of accurately reproducing the ions in the GCR spectrum [13].

So what other metrics can be applied to simulate the effects of GCRs? Well, one way is to apply the metric of linear energy transfer, or LET (see boxed material). LET is used to measure the amount of tissue damage caused by radiation and to determine radiation protection and risk assessment. This metric is used in conjunction with the measure of relative biological effectiveness (RBE), which is applied to the effect of different types of radiation. In essence, the higher the RBE for a

Fig. 6.7 Orion. (Image courtesy of NASA)

specific type of radiation, the more damaging that radiation is per unit of energy when it is absorbed by human tissues. Several studies utilizing LET and RBE have been conducted over the years by measuring the LET spectrums using tissue equivalent proportional counters, and plastic nuclear track detectors placed at different locations on board the ISS [14, 15].

Similar studies were conducted during NASA's Exploration Flight Test (EFT-1), which studied the Orion (Fig. 6.7) Multi-Purpose Crew Vehicle (MPCV) during orbital flight. Although the 4-h EFT-1 flight was much shorter than 6-month ISS flights, the high apogee (5800 km) of the second orbit included a transit through the radiation-dense Van Allen Belts and also a brief excursion into the interplanetary environment. Radiation detectors were activated shortly after lift-off and collected radiation data for the duration of the flight.

Spallation

When highly charged particles penetrate shielding or astronauts, they do so in a straight path to begin with. But shortly after penetrating matter, those heavy ions begin to disperse as they collide with atoms in the shielding and/or the astronauts. As the path of the heavy ions is disrupted, energy is dissipated, but at the same time smaller nuclei are generated in a process called *spallation*. The degree to which energy is dissipated is largely determined by the properties of the material through which the heavy particles are traveling. Generally, energy loss increases with decreasing atomic number, which is why hydrogen is such an effective shielding material. Scientists can calculate the stopping power of a material by determining the energy lost per unit path length that the particles travel. This number is the LET, which is a metric that quantifies how much energy is lost as the heavy ions transit material. But stopping power isn't everything when it comes to choosing a shielding material. A good shielding material should not only stop as many of the high-energy particles as possible but also limit the amount of fragmentation as much as possible. Polymers tend to be good candidate shield materials because they have a high hydrogen content and also stop more low-energy particles than most other materials. But the choice of the material is just one consideration. The next decision is how thick the material should be. This is important because the LET of the heaviest nuclei has such penetrating energy that they can travel deep through a material before any measurable energy loss. It is therefore important that the shielding material is designed in such a way that spallation is limited and energy loss is maximized. This is very difficult to do because the data on fluence of particles in deep space is limited (one application that is used to calculate this is the Monte Carlo particle transport simulation software PHITS). Even with advanced simulation software such as PHITS, reliable and accurate predictions of how well a shield will function is very difficult. This is because of the lack of data from deep space and the difficulty in predicting how neutron propagation (which is highly sporadic) occurs in biological tissue.

Utility and Applications of LET

Today, heavy-ion accelerators are used to expose animals to the entire projected radiation dose for an interplanetary mission. But these single-ion and heavy-ion exposures do not accurately simulate the highly complex dose rate or the energetic dispersion of GCRs, and therefore do not accurately model the multi-organ toxicity of such radiation.

So what can scientists do to replicate the effect of GCRs in a ground-based analog? Well, NASA has suggested building a GCR simulator that will generate more LET data points, but that will only be one step of several that will be needed. Radiation engineer Pellish says: "Ultimately, the solution to radiation will have to be a combination of things. Some of the solutions are technology we have already, like hydrogen-rich materials, but some of it will necessarily be cutting edge concepts that we haven't even thought of yet."

Polyethylene as a Shielding Material

One candidate for radiation shielding is polyethylene (Fig. 6.8), a plastic that is also found in water bottles. By virtue of its high hydrogen content [8] and the fact that it is very cheap, this material also offers other advantages when it comes to protecting

Fig. 6.8 Dr. Raj Kaul examines "bricks" of radiation shielding material made in his composites laboratory at NASA's Marshall Space Flight Center in Huntsville, AL. Kaul has investigated the effectiveness of material used to shield spacecraft from radiation. (Image courtesy of NASA/MSFC/Doug Stoffer)

astronauts from radiation. For example, plastic-like materials such as polyethylene cause much less secondary radiation than traditional materials such as aluminum. Since polyethylene isn't the most versatile material when it comes to building spacecraft, it has been adapted to a stronger and lighter material: RXF1. Created by Raj Kaul, RXF1 is derived from polyethyene, but since it is a fabric it can be shaped into whatever is needed.

Hydrogenated Boron Nitride Nanotubes

Although polyethylene has been shown to be effective at dispersing heavy ions, stopping protons, and slowing down neutrons (which form as secondary radiation), it is not a structural material, although RXF1 may have some potential in that application. But there is another hydrogen-based material that might do both jobs [16–18]. Hydrogenated boron nitride nanotubes (Fig. 6.9), also known as hydrogenated BNNTs, comprise nanotubes constructed of carbon, boron, nitrogen and hydrogen. In addition to absorbing secondary neutrons and stopping protons, the hydrogenated BNNT material is so flexible that spacesuits could be made of it. Oren Milstein, CEO of StemRad, says, "This product will enable human deep space exploration. Our breakthrough has come in creating the architecture of the multi-layered shield to accurately cover the most important organs."

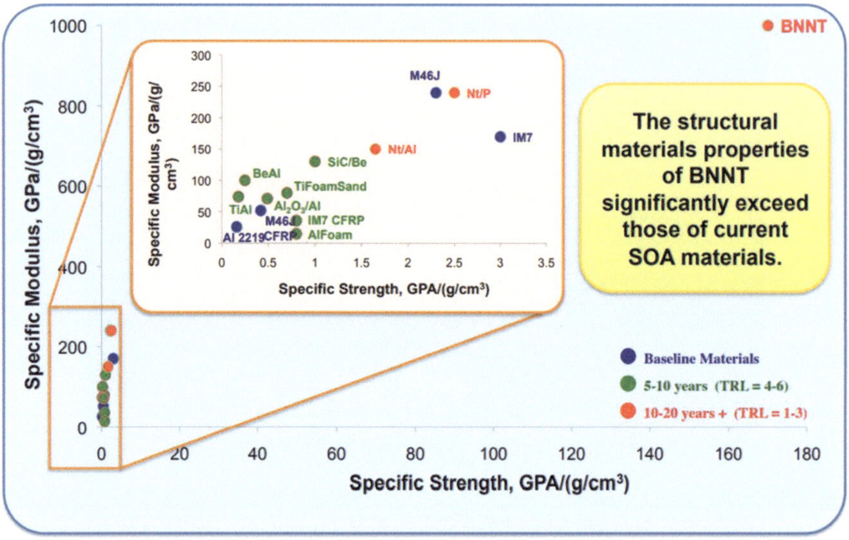

Fig. 6.9 BNNT. (Image courtesy of NASA) (see also Appendix C)

Fig. 6.10 Stemrad's radiation shielding concept—the Astrorad. (Image courtesy of NASA)

AstroRad Radiation Shield

Another passive means of protecting astronauts is a vest being developed by Israeli researchers. Dubbed the AstroRad Radiation Shield, the vest is being produced by StemRad, a company based in Tel Aviv. The vest (Fig. 6.10), which will be customized for each crewmember, is designed to protect vital organs. To test the concept, the vest will be 'worn' by a phantom torso that will measure radiation absorption. Another phantom torso will be flown that will be unprotected.

References

1. Chancellor JC, Scott GB, Sutton JP. Space radiation: the number one risk to astronaut health beyond low Earth orbit. Life. 2014;4:491.
2. Kennedy AR. Biological effects of space radiation and development of effective countermeasures. Life Sci Space Res. 2014;1:1.
3. Romero-Weaver AL, Wan XS, Diffenderfer ES, Lin L, Kennedy AR. Effect of SPE-like proton or photon radiation on the kinetics of mouse peripheral blood cells and radiation biological effectiveness determinations. Astrobiology. 2014;13:570.
4. Wilson JW, Thibeault RC, Cucinotta FA, Shinn ML, Kim MH, Kiefer R, Badavi FF. Issues in protection from galactic cosmic rays. Radiat Environ Biophys. 1995;34:217.
5. Walker SA, Townsend LW, Norbury JW. Heavy ion contributions to organ dose equivalent for the 1977 galactic cosmic ray spectrum. Adv Space Res. 2013;51:1792.
6. Durante M, Cucinotta FA. Heavy ion carcinogenesis and human space exploration. Nat Rev Cancer. 2008;8:465.
7. Sternheimer RM. Density effect for the ionizing loss of charged particles in various substances. Phys Rev B. 1982;26:6067.
8. Zeitlin C, Heilbronn CL, Miller J, Schimmerling W, Townsend LW, Tripathi RK, Wilson JW. The fragmentation of 510 MeV/nucleon iron-56 in polyethylene. II. Comparisons between data and a model. Radiat Res. 1996;145:666–72.
9. Durante M, George K, Gialanella G, Grossi G, La Tessa C, Manti L, Miller J, Pugliese M, Scampoli P, Cucinotta FA. Cytogenetic effects of high-energy iron ions: dependence on shielding thickness and material. Radiat Res. 2005;164:571–6.
10. Zeitlin C, Guetersloh S, Heilbronn L, Miller J. Shielding and fragmentation studies. Radiat Prot Dosim. 2005;116:123–4.
11. Badhwar GD, O'Neill PM. An improved model of GCR for space exploration missions. Nucl Tracks Radiat Meas. 1992;20:403–10.
12. Aiginger H, et al. The FLUKA code: new developments and application to 1 GeV/n iron beams. Adv Space Res. 2005;35:214–22.
13. Armstrong TW, Chandler KC. Stopping powers and ranges for muons, charged pions, protons, and heavy ions. Nucl Inst Methods. 1973;113:313.
14. Shavers MR, Zapp N, Barber RE, Wilson JW, Qualls G, Toupes L, Ramsey S, Vinci V, Smith G, Cucinotta FA. Implementation of ALARA radiation protection on the ISS through polyethylene shielding augmentation of the Service Module Crew Quarters. Adv Space Res. 2004;34:1333–7.
15. Stoffle N, Pinsky L, Kroupa M, Hoang S, Idarraga J, Amberboy C, Rios R, Hauss J, Keller J, Bahadori A, et al. Timepixbased radiation environment monitor measurements aboard the International Space Station. Nucl Instrum Methods Phys Res Sect A. 2015;782:143.
16. Kim, MHY, Wilson, JW, Thibeault SA, Nealy JE, Badavi FF, Kiefer, RL. Performance study of galactic cosmic ray shield materials. NASA-TP-3473, 1994.
17. Harrison C, Weaver S, Bertelsen C, Burgett E, Hertel N, Grulke E. Polyethylene/boron nitride composites for space radiation shielding. J Appl Polym Sci. 2008;109:2529–38.
18. Estevez JE, Ghazizadeh M, Ryan JG, Kelkar AD. Simulation of hydrogenated boron nitride nanotubes mechanical properties for radiation shielding applications. Int J Chem Sci Eng. 2014;8(1):63–7.

Chapter 7
Acute Radiation Sickness

The risk of a crew suffering from acute radiation syndrome (ARS) has been identified as a significant threat during future exploration class missions [1, 2] (Fig. 7.1). The syndrome includes symptoms that occur within hours to days following exposure to radiation (Table 7.1).

ARS is comprised of several health effects that occur within 24 h of being exposed to high levels of ionizing radiation [3]. Astronauts being exposed to the levels that result in ARS will suffer profound and potentially deadly symptoms, as cellular degradation occurs due to damage to DNA and other cell structures. Depending on the severity of the dose, symptoms may present within one hour of exposure. If the exposure is relatively mild, symptoms will include nausea, vomiting, falling blood counts and an increased susceptibility to infection. At this level of exposure the crew will probably recover, and the mission will remain intact. But if crewmembers are exposed to high levels of exposure symptoms may include neurological effects and ultimately death (Table 7.2). Another scenario is if the crew is exposed to several months or years of moderate levels of radiation, in which case there is the risk of succumbing to chronic radiation syndrome [4, 5] (Fig. 7.2).

Signs and Symptoms

ARS (Fig. 7.3) is divided into three syndromes, each of which is outlined here. In some cases these syndromes may be preceded by a *prodrome*, which is an early sign that indicates the onset of a syndrome of disease. In each syndrome the rate at which symptoms present is a function of radiation exposure—generally, the greater the dose, the shorter the delay between onset of symptoms. Also, for each syndrome to occur, the tissues specific to the body system must be exposed. For example, for the gastrointestinal syndrome to occur, the tissues of the gastrointestinal system must have been exposed to radiation.

E. Seedhouse, *Space Radiation and Astronaut Safety*, SpringerBriefs in Space Development, https://doi.org/10.1007/978-3-319-74615-9_7

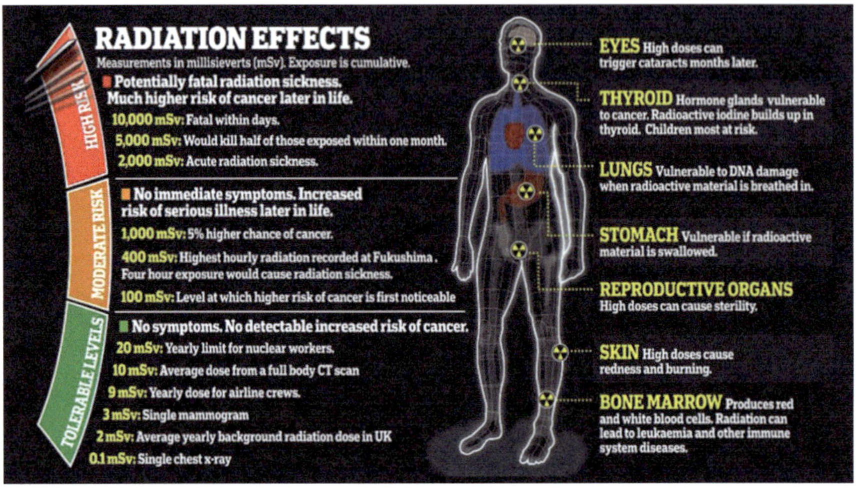

Fig. 7.1 Radiation effects (Image courtesy of Go Flight Medicine)

Table 7.1 Acute radiation sickness symptoms and mortality[a]

		Whole-body absorbed dose (in Gy)				
Phase	Symptoms	1–2 Gy	2–6 Gy	6–8 Gy	8–30 Gy	>30 Gy
Immediate	Nausea/vomiting	5–50%	50–100%	75–100%	90–100%	100%
	Time of onset	2–6 h	1–2 h	10–60 mins	<10 mins	Minutes
	Duration	<24 h	24–48 h	<48 h	<48 h	Death in <48 h
	Diarrhea	None	None to mild	Heavy	Heavy (>95%)	100%
	Time of onset		3–8 h	1–3 h	<1 h	<1 h
	Headache	Slight	Mild to moderate	Moderate	Severe	100%
	Time of onset		4–24 h	3–4 h	1–2 h	<1 h
	Fever	None	Moderate increase	Moderate to severe	Severe	100%
	Time of onset		1–3	<1 h	<1 h	1 h
	CNS Function	No impairment	Cognitive impairment 6–20 h	Cognitive impairment >24 h	Rapid incapacitation	Seizures, tremors, ataxia
Latent period						
Illness		Mild to moderate leukopenia, fatigue	Moderate to severe leukopenia, infections Alopecia	Severe leukopenia Fever Vomiting Dizziness	Nausea Vomiting High fever Electrolyte imbalance	Patients die in less than 48 h
Mortality	Without care (%)	0–5	5–95	95–100	99–100	100
	With care (%)	0–5	5–50	50–100	100	100
	Death	6–8 weeks	4–6 weeks	2–4 weeks	2 days to 2 weeks	1–2 days

[a]Merck Manual. Radiation Exposure and Contamination, by Jerrold T. Bushberg, PhD, DABMP, Clinical Professor, Radiology and Radiation Oncology, and Director of Health Physics Program, School of Medicine, University of California, Davis

Table 7.2 Required conditions for acute radiation syndrome (ARS)[a]

1. The radiation dose must be large (i.e., greater than 0.7 Gy)
2. Mild symptoms may be observed with doses as low as 0.3 Gy
3. The dose usually must be external
4. The radiation must be penetrating (i.e., able to reach the internal organs)
5. High energy X-rays, gamma rays, and neutrons are penetrating radiations
6. The entire body (or a significant portion of it) must have received the dose
7. The dose must have been delivered in a short time (usually a matter of minutes)
8. Fractionated doses are often used in radiation therapy. These are large total doses delivered in small daily amounts over a period of time

[a]Adapted from: Centers for Disease Control and Prevention http://emergency.cdc.gov/radiation/ars.asp

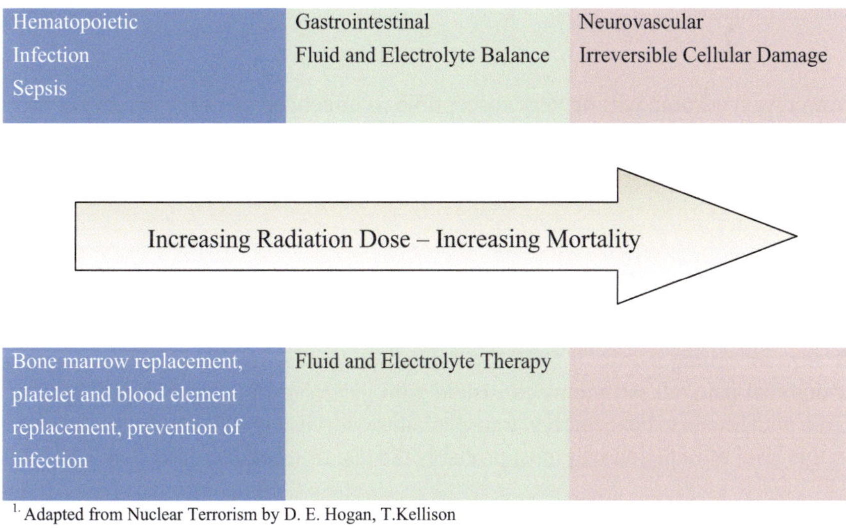

Fig. 7.2 Spectrum of radiation sickness. Adapted from Nuclear Terrorism by D. E. Hogan, T. Kellison

Hematopoietic Syndrome

A part of ARS is the hematopoietic syndrome (Fig. 7.4) that results from radiation doses that may be encountered during exposure to a solar particle event (SPE). The hematopoietic syndrome is characterized by a reduction in the number of blood cells, a process that occurs as a result of radiation-induced cell killing. An excess reduction in the number of blood cells may be accompanied by a reduction in stem cells in the bone marrow that may ultimately lead to bone marrow failure [6, 7]. The magnitude of blood cell counts is one means of assessing radiation dose, which in turn can be used to determine therapy and a prognosis. A crewmember suffering

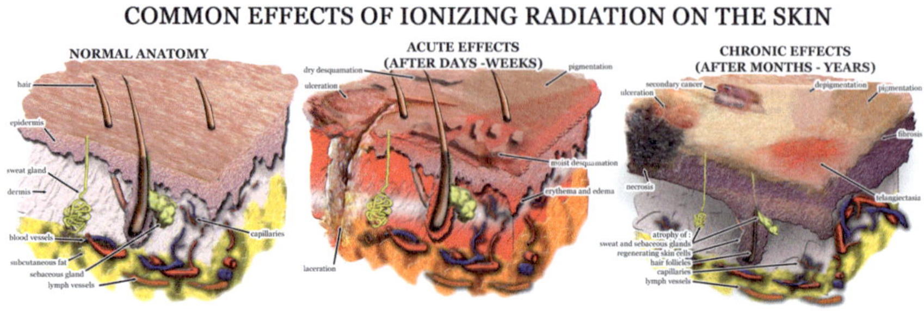

Fig. 7.3 High doses of radiation can cause browning of skin, known as a "nuclear tan". If radiation exposure is particularly high, the result is first or second degree burns that may be accompanied by dry or wet desquamation. If the radiation very intense, third of fourth degree burns may occur, accompanied by wet ulcerated lesions and eventual necrotic dermatitis. (Image courtesy of Virinchi Hospitals)

from this syndrome will be very susceptible to infections due to a significant drop in white blood cells. The problem will be exacerbated by bleeding that will result in a reduction in platelets and consequent anemia.

Gastrointestinal Syndrome

A crew absorbing a dose of 6–30 Gy will suffer nausea, vomiting and excruciating abdominal pain, all symptoms consistent with gastrointestinal syndrome (Fig. 7.4). Even with access to bone marrow transplantation, survival is unlikely. A crew exposed to this level of radiation will most probably die due to secondary infection [8].

Neurovascular Syndrome

Symptoms of this syndrome (Fig. 7.4) occur in those exposed to more than 30 Gy, although in susceptible individuals 10 Gy may be enough to trigger symptoms. At this very high level of exposure symptoms such as dizziness and unconsciousness present within minutes or hours. Death follows shortly thereafter [8].

Genetic Damage

The high radiation doses that cause ARS also cause damage to DNA, and if DNA cannot be repaired chromosomal abnormalities occur. The process of DNA damage follows a set sequence of events. First, radiation causes clusters of DNA damage, which results in the loss of nucleobases. Nucleobases are biological compounds that

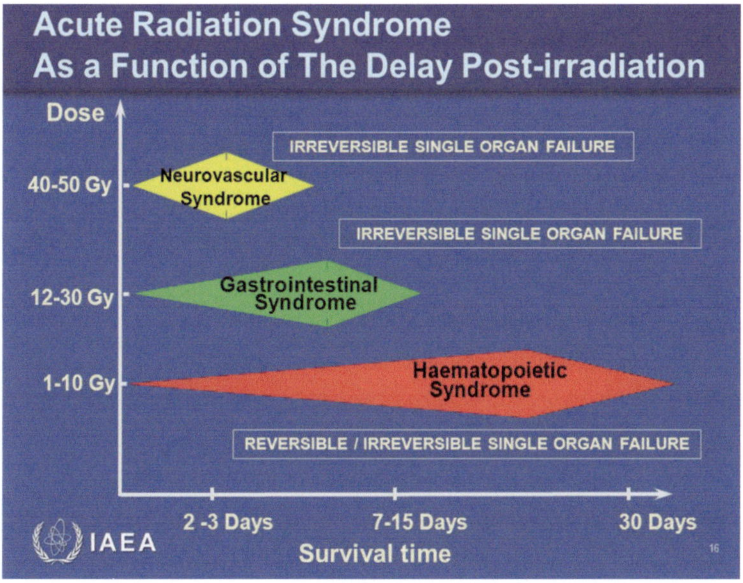

Fig. 7.4 Effects of ARS. (Image courtesy of IAEA)

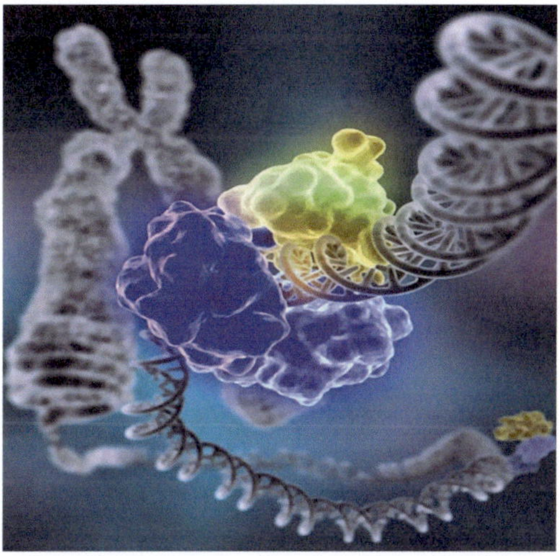

Fig. 7.5 High radiation doses will inevitably lead to genetic damage. (Image courtesy of NASA)

form nucleosides, which are a component part of nucleotides. Together, nucleobases, nucleosides and nucleotides form the building blocks of nucleic acids, and it is the ability of the nucleobases to form base pairs that result in the helical structures that are ribonucleic acid and deoxyribonucleic acid—DNA. But if the nucleobases cannot form base pairs, these helical structures cannot form, and the backbone of the genetic structure is compromised [2]. Although single-strand breaks can be relatively easy to repair, double-strand breaks (Fig. 7.5) are more difficult to put back together.

Although repair processes within DNA occur naturally, it is the clustered damage unique to radiation that is especially harmful (see boxed text). That is because clustered damage results in lesions, and this type of damage takes much longer to repair than isolated breaks. And the greater and more intense the radiation the more extensive the number of lesions and the less likely this damage is repairable. And when the damage cannot be repaired the result is usually a phenotype manifested as a mutation of death of a cell or cells. Although somatic mutations will not be passed on from a parent, these types of mutations do cause alterations within cells and, depending on what stage that cell is in its mitotic cycle, the alterations may be replicated, resulting in a cascade effect that may ultimately result in cell death [3].

Predicting Acute Radiation Symptoms
To predict acute radiation symptoms, the *whole-body absorbed dose* is used, and this metric is measured in Gy[1] (the SI unit symbol is Gy). But what does this mean? Here is an example. Imagine if a Mars-bound crew was exposed to a major solar flare event and a crewmember was subjected to 10 Gy. If 25% of this astronaut's bone marrow mass was irradiated then the absorbed dose for the bone marrow would be 2.5 Gy. But, because bone marrow is comprised of just four percent of the body mass, this astronaut's whole-body dose would be 0.25 Gy.

Management

There have been no controlled studies of treating humans for ARS. Treatments and recommendations such as the administration of antibiotics, blood products, and stem cell transplant are based on animal research, but only the first of these treatments would be available on an interplanetary spacecraft.

Deterministic and Stochastic Effects

In the event of the crew being exposed to a radiation event the crew medical officer (CMO) would determine the lethality of the radiation exposure. This would be calculated by measuring the dose rate and the distance from the source of radiation. From this information an estimation of the lethal dose (LD) at 60 days could be made. For example, the LD50/60 is the dose of radiation required to result in death in sixty days. Without supportive care, the LD at sixty days is about 4 Gy, between 4.5 and 7 Gy when antibiotics and treatment is available, and between 7 and 9 Gy for patients with immediate access to intensive care units and cell transplants. For exposures exceeding 10 Gy, no survival is possible [9, 10]. As part of the assessment of radiation exposure, the CMO would calculate the *deterministic* effect and the *stochastic* effect.

[1] The Gray (G) is the absorption of one joule of radiation per kilogram of matter.

Table 7.3 Deterministic effects

Effects	Dose (Gy)
Skin erythema	2–5[a]
Irreversible skin damage	20–40
Sterility	2–3
Cataracts	5
Lethality	3–5

[a]1 rad = 0.01 Gy. 1 rem = 0.01 Sv

Deterministic effects, which are also referred to as *non-stochastic* effects, are the cause and effect associations between radiation and specific side effects. The threshold for deterministic effects may vary from person to person; for some people, the threshold may be low and for others the threshold may be high. Some examples of deterministic effects given as the absorbed dose are presented in Table 7.3.

Stochastic effects represent the outcome of exposure to radiation. For example, one stochastic effect is radiation-induced carcinogenesis. To calculate an individual's stochastic risk it is necessary to apply the radiation dose to the individual's age. But there is no clear linear relationship between the amount of the dose and the severity of the effect. So, while the risk of cancer increases with the dose of radiation, the outcome of being exposed to that dose has no bearing on the severity of the effects; the exposed individual will either develop cancer or not.

Stages of Acute Radiation Syndrome

An estimate of an individual's dose can be determined by the use of body dosimeters (see Chap. 5). The *dose* is calculated as the whole-body dose or a significant partial-body irradiation, but the *symptoms* of ARS are related to the whole-body dose. Acute symptoms are expected following exposures to doses greater than 1 Gray if these doses have been delivered in a short period of time (minutes). This dose will then determine the symptoms for each of the four stages of ARS.

Prodromal Stage

The signature symptoms for this stage include nausea, vomiting and diarrhea. These symptoms will occur from minutes to days after the exposure [9–11]. Symptoms may last for several days. If an individual has been exposed to a lethal dose exceeding 2 Gy, symptom onset will occur within two hours, whereas if the dose is between 10 to 20 Gy, symptoms will occur within minutes.

Latent Stage

In this stage, patients will appear healthy for a period that may last a few hours or may extend to a few weeks.

In the manifest illness stage, symptoms will be syndrome specific (Table 7.4) and will last for hours or months. Recovery may take several weeks, months or extend to up to 2 years. Those who do not recover typically die within months of exposure.

Table 7.4 Radiation effects

Syndrome	Dose	Prodromal stage	Latent stage	Manifest illness stage	Recovery
Hematopoietic (bone marrow)	>0.7 Gy (mild symptoms may occur as low as 0.3 Gy)	Symptoms are anorexia, nausea and vomiting. Onset occurs 1 h to 2 days after exposure Stage lasts for minutes to days	Stem cells in bone marrow die, although patient may appear well Stage lasts 1–6 weeks	Symptoms are anorexia and fever Drop in blood cell counts over several weeks Primary cause of death is infection and hemorrhage. Survival decreases with increasing dose	In most cases, bone marrow cells begin to repopulate the marrow Should be full recovery for most individuals from a few weeks up to 2 year after exposure Death may occur in some individuals at 1.2 Gy LD50/60 is about 2.5–5 Gy
Gastrointestinal (GI)	>10 Gy (some symptoms may occur as low as 6 Gy)	Symptoms are anorexia, severe nausea, vomiting, and diarrhea Onset occurs within a few hour after exposure Stage lasts about 2 day	Stem cells in bone marrow and cells lining GI tract die. Patient may feel well Stage lasts <1 week	Symptoms are malaise, anorexia, severe diarrhea, fever, and electrolyte imbalance Death is due to infection, & electrolyte imbalance Death occurs within 2 week of exposure	LD100 is about 10 Gy
Cardiovascular (CV)/central nervous system (CNS)	>50 Gy (some symptoms may occur as low as 20 Gy)	Symptoms are extreme confusion, severe nausea, vomiting, loss of consciousness Onset occurs within mins of exposure Stage lasts for minutes to hours	Patient may return to partial functionality Stage may last for hours	Symptoms are return of diarrhea, convulsions, and coma Onset occurs 5–6 h after exposure Death occurs within 3 days of exposure	No recovery expected

Dealing with ARS in Deep Space

ARS is comprised of a series of severe injuries that occur after the whole body or a significantly large part of the body has been exposed to a high dose of radiation. It is a type of injury that affects all organs to some degree, although the extent of the injury depends on the intensity and magnitude of the ionizing radiation. How an individual responds to such an exposure will depend to some degree on that person's radiosensitivity. But regardless of an individual's sensitivity, a crew being exposed to a high dose (see Appendices 1 and 2 of this book) of ionizing radiation will be subject to the clinical components of ARS with its various syndromes. Management of these syndromes, which is discussed in Chap. 8, will be *far* beyond the capabilities of any spacecraft currently being designed for interplanetary missions. A crew suffering from intermediate to high-dose radiation (between 4 and 10 Gy) will need urgent hematopoietic stem cell transplantation (HSCT), a procedure that is difficult enough in an advanced health care facility, never mind inside the claustrophobic confines of a spacecraft (Fig. 7.6).

Fig. 7.6 Yuri Gidzenko in the ISS Leonardo Module. (Image courtesy of NASA)

References

1. Hellweg CE, Baumstark-Khan C. Getting ready for the manned mission to Mars: the astronauts' risk from space radiation. Naturwissenschaften. 2007;94(7):517–26.
2. Townsend LW. Implications of the space radiation environment for human exploration in deep space. Radiat Prot Dosimetry. 2005;115(1-4):44–50.
3. Hall EJ. Acute effects of total-body irradiation. In: Hall EJ, editor. Radiobiology for the radiologist. 5th ed. Philadelphia: Lippincott Williams and Wilkins; 2000. p. 117–28.
4. Anno GH, Baum SJ, Withers HR, Young RW. Symptomatology of acute radiation effects in humans after exposure to doses of 0.5-30 Gy. Health Phys. 1989;56(6):821–38.
5. Cerveny TJ, MacVittie TJ, Young RW. Acute radiation syndrome in humans. In: Walker RI, Cerveny TJ, editors. Medical consequences of nuclear warfare 1989. Falls Church: TMM Publications, Office of the Surgeon General; 1989. p. 17–36.
6. Donnelly EH, Nemhauser JB, Smith JM, et al. Acute radiation syndrome: assessment and management. South Med J. 2010;103(6):541–6.
7. Berger ME, O'Hare FM Jr, Ricks RC, editors. The medical basis for radiation accident preparedness: the clinical care of victims. REAC/TS conference on the medical basis for radiation accident preparedness. New York: Parthenon Publishing; 2002.
8. Gusev IA, Guskova AK, Mettler FA Jr, editors. Medical management of radiation accidents. 2nd ed. New York: CRC Press, Inc.; 2001.
9. Waselenko JK, MacVittie TJ, Blakely WF, et al. Medical management of the acute radiation syndrome: recommendations of the Strategic National Stockpile Radiation Working Group. Ann Intern Med. 2004;140(12):1037–51.
10. Weisdorf D, Chao N, Waselenko JK, et al. Acute radiation injury: contingency planning for triage, supportive care, and transplantation. Biol Blood Marrow Transplant. 2006;12(6):672–82.
11. Jarrett DG. Medical management of radiological casualties handbook. 1st ed. Bethesda: Armed Forces Radiobiology Research Institute (AFRRI); 1999.

Chapter 8
Treating Acute Radiation Syndrome

A crew exposed to a high dose of ionizing radiation will fall into three categories: (1) those who recover with minor medical intervention, (2) those who require high-level medical care (such as bone marrow stem cell transplantation), and (3) those triaged for palliative care [1, 2]. In reality, given the lack of medical supplies, the second group will fall into the third category. In such an event (see boxed text), the crew medical officer (CMO) will need to remove external contamination, estimate the radiation dose (Table 8.1), and administer fluids and electrolytes (Figs. 8.1 and 8.2). That is assuming the CMO is not one of those who falls into the second or third category! After 48 h, a second patient scoring will be conducted, and documentation of therapeutic management and organ failure will be performed.

The K-19 Incident

K-19 (Fig. 8.3) was a first generation Russian nuclear Hotel-class submarine equipped with R-13 submarine-launched ballistic missiles. Although not a spacecraft, life on board a nuclear submarine is often used as an analog of life on board a spacecraft. In an effort to catch up with the U. S. lead in nuclear submarine technology, K-19 was built in a rush and was plagued with mishaps. Ten workers died on the ship before the submarine was even launched. Then, on its maiden voyage in July 1961, K-19 narrowly avoided a nuclear meltdown when the coolant system failed. Predictably, the reactor temperature rose uncontrollably, forcing Zatayev, the commander, to order his crew to construct a makeshift coolant system.

Radiation conditions continued to deteriorate. I summoned my starpom, Captain Lieutenant Yenin, to the cockpit and authorized him to give the entire crew 100-g servings of liquor. I knew that under the influence of alcohol (which is also a narcotic) the cells of the body are less susceptible to radiation; in other words, resistance to external irritants goes up... On the strength of my experience of an accident in Obninsk, when an operator had dosed himself up on liquor before going to work on a reactor core and radiation had not affected him, I authorized this measure. (K-19, p. 134.)

E. Seedhouse, *Space Radiation and Astronaut Safety*, SpringerBriefs in Space Development, https://doi.org/10.1007/978-3-319-74615-9_8

Table 8.1 ARS sub-syndromes at dose thresholds

Radiation dose	Signs and symptoms	Time: exposure to presentation
5 rem	Chromosome aberrations first seen	30 min
12 rem	Reduction in sperm count	42 days
0.75 Gy	Lymphocyte depletion	6 h
1 Gy	Nausea, vomiting	6 h, then 5–7 day
1–6 Gy	Hematopoietic syndrome	1–6 h
3 Gy	Temporary epilation	14 days
6 Gy	Erythema	6–48 h, then 2–3 week
6 Gy	Pneumonitis	4–6 week
6 Gy	Pulmonary syndrome	1–6 month
6–8 Gy	Gastrointestinal syndrome	3–4 day
9–10 Gy	Death	Days to weeks
>10 Gy	Neurovascular syndrome	Hours to days

Phases of Radiation Injury

Dose (Gy)	Prodromal Phase	Manifest Phase	Prognosis without Supportive Care
0.5–1.0	Mild	Modest decline in blood counts	Survival
1.0–2.0	Mild–moderate	Some bone marrow damage	Survival >90%
2.0–3.5	Moderate	Moderate–severe bone marrow damage	Probable survival
3.5–5.5	Severe	Severe bone marrow damage; modest GI damage	Death within 3.5–6 wk (50% of victims)
5.5–7.5	Severe	Pancytopenia and moderate GI damage	Death probable within 2–3 wk
7.5–10.0	Severe	Severe GI and bone marrow damage	Death probable within 2 wk
10	Severe	Severe GI damage, radiation-induced lung injury, altered mental status; at higher doses (>20.0 Gy), cardiovascular collapse, fever, shock	Death within 2 wk

Ann Intern Med. 2004;140:1037–1051.

Fig. 8.1 Phases of radiation injury (Image courtesy of the Annals of Internal Medicine)

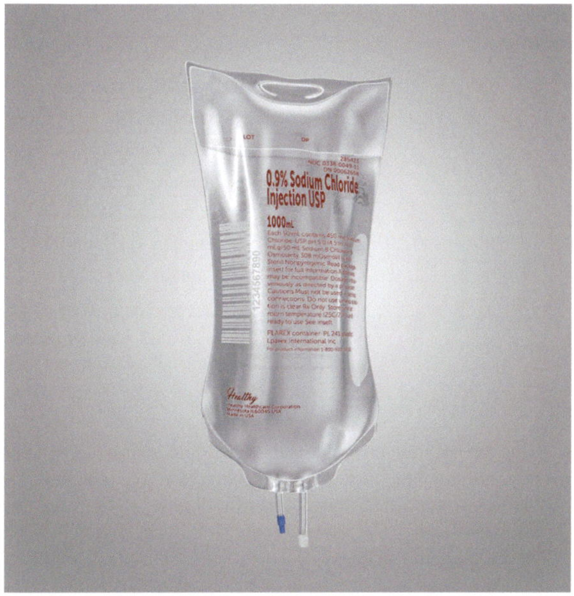

Fig. 8.2 Saline solution. (Image courtesy of Healthy Healthcare Corporation)

Fig. 8.3 K-19. (Open source image. Military Clips)

Table 8.2 K-19 crewmember fatalities

Name	Sv	Date of death
Lt B. Korchilov	54	10 July 1961
Chief B. Ryzhikov	8.6	25 July 1961
Starshina 1st class Y. Ordochkin	11	10 July 1961
Starshina 2nd class, E. Kashenkov	10	10 July 1961
Seaman S. Penkov	10	18 July 1961
Seaman N. Savkin	11	13 July 1961
Seaman V. Charitonov	11	15 July 1961
Captain Y. Povstyev	7.5	22 July 1961

[a]Sievert is equal to 100 rem (roentgen equivalent man) K-19 The widowmaker: The Secret Story of the Soviet Nuclear Submarine. Huchthausen P

This job required crewmembers to work in a high radiation environment for extended periods. Although the makeshift system reduced the temperature in the reactor, the entire crew had been exposed to high levels of radiation, eight of whom sustained doses of more than 5000 rems. All eight crewmembers (Table 8.2) died within three weeks of the incident from severe radiation sickness.

Treating Hematopoietic Syndrome

This ARS syndrome is normally seen at doses in excess of 1 Gy [3]. Classic signs and symptoms are usually observed after one of two weeks following exposure, although evidence of damage can be seen in the blood cell count after just a few days. For example, lymphocytes decrease within 12–24 h following exposure, and with higher doses, within 6–12 h. The rate at which lymphocytes (Fig. 8.4) decrease is helpful to the physician, as it can help guide treatment [4].

In terms of severity, mild injury is considered to occur when a patient has been exposed to between 1 and 2 Gy, moderate injury between 2 and 4 Gy, and severe injury between 4 to 6 Gy. Very severe injury occurs beyond 6 Gy. For the CMO, treatment will be partly determined by lymphocyte depletion kinetics as depicted in the Andrews curves in Fig. 8.4.

Since red blood cells (RBCs) are more resistant to radiation than the precursor cells, there will be no requirement for blood infusions unless this is needed to support oxygen delivery due to blood loss (from other trauma, for example) [5, 6]. But, if the crewmember was exposed to more than 2 Gy, administration of prophylactic antimicrobial drugs may be required. And in the case of a crewmember suffering profound neutropenia, stem cell transplantation will be the only option [7, 8]. But, since this option will not be supported by the medical capabilities on board an interplanetary spacecraft, palliative care will be the only option.[1]

[1]In reality, palliative care will not be an option. Without stem cell transplantation, a crewmember will not survive. Rather than use life support consumables, the only viable option for this crewmember would be voluntary euthanization.

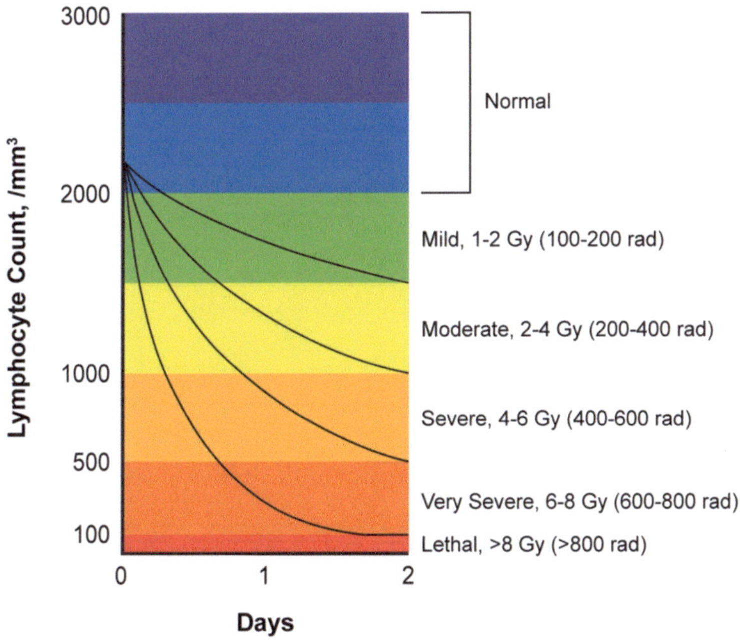

Fig. 8.4 Andrews curves. These curves describe the pattern of lymphocyte response in relation to dose. (Image courtesy of NASA)

In crewmembers who have suffered high levels of ionizing radiation, hematopoietic stem cell transplantation (HSCT) is the only possible medical intervention. On Earth, HCST (Fig. 8.5) is complicated by factors such as radiation-induced bone marrow failure, which determines the homogeneity of radiation exposure (some marrow-containing parts of the body might have been minimally irradiated thanks to shielding). Physicians making the call on whether to commence HCST also have to consider myriad other factors such as antibiotics and antivirals, barrier isolation, cytokine therapy, wound closure, isolation rooms, and rigorous environmental control. None of these considerations can be managed on an interplanetary spaceship.

At best, the interplanetary CMO might be able to offer meaningful treatment to those crewmembers who have suffered low levels of exposure (<1 Gy), but beyond this, treatments are the preserve of fully-fledged medical facilities. For example, erythropoietin (EPO) anemia therapy is a candidate therapy for those who have been exposed to 2 or 3 Gy, but this therapy can only be performed in a hospital setting. Similarly, granulocyte colony-stimulating factor (G-CSF) and cytokine therapy[2] are viable treatments [9] for those who have received a whole-body dose of 3 Gy, but such treatments could not be administered with the limited medical facilities on board our interplanetary spacecraft.

[2] Patients diagnosed with severe neutropenia would be candidates for being treated with the following cytokines: Filgrastim at 2.5–5 μg/kg/day subcutaneously; Sargramostim at 5–10 μg/kg/day subcutaneously, and Pegfilgrastim at 6 mg subcutaneously.

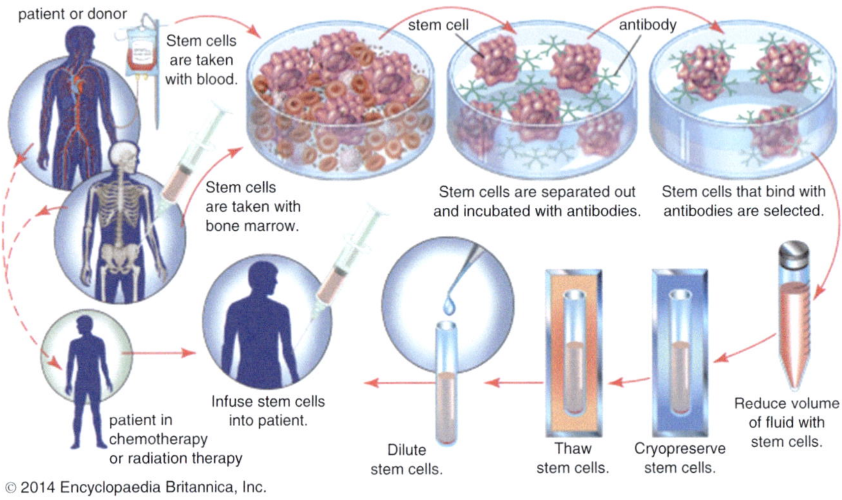

Fig. 8.5 Human stem cell transplantation. (Open source image from the Encyclopedia Britannica)

Treating Gastrointestinal Syndrome

Classic signs and symptoms of this ARS syndrome, which include nausea and vomiting, are usually seen at doses in excess of 5 Gy [10]. These signs and symptoms are caused primarily by the disruption of the mechanism of replacing intestinal epithelial stem cells found in the gastrointestinal microvilli. Under normal conditions these stem cells mature and are replaced, but following exposure or ionizing radiation, disorganization occurs and this results in malabsorption, dehydration, bleeding, fluid and electrolyte imbalance, and renal failure. In common with the hematopoietic syndrome, the gastrointestinal syndrome is characterized by faster onset of symptoms the higher the dose of ionizing radiation. Treatment may include administration of fluoroquinolone two to four days after exposure, digestive decontamination, and, for exposures greater than 2 Gy, enteral nutrition and proton pump inhibitors [11].

Treating Neurovascular Syndrome

The signs and symptoms associated with this syndrome begin early and at low exposure levels, although the classic signs and symptoms are observed at exposures greater than 10 Gy. As with the previous two syndromes, the severity of signs and symptoms is dependent on dose. Those who have been exposed to 10 Gy or more generally don't survive more than a few days following exposure. Symptoms include nausea, vomiting, lethargy, cognitive dysfunction, ataxia, seizures, cerebral edema, and hypotension, although those who have suffered doses higher than 10 Gy may die before all signs and symptoms present. Treatment includes mannitol, furosemide, and analgesics [11].

General Treatment

One of the general risks to patients following acute exposure to ionizing radiation is increased susceptibility to local and systemic infection. To reduce this risk, treatments include administration of antivirals such as fluoroquinolone [1]. For those patients suffering fever and neutropenia, the administration of broad spectrum prophylactic antimicrobials is a standard treatment that, if the neutropenic duration is extended, can be combined with intravenous monotherapy with meropenem or imipenem. Given the extremely limited medical resources on board it is unlikely these treatment options will be available to the crew.

Chief Medical Officer's Response to Radiologic Incident

The CMO's first task will be to estimate the radiation dose, as it is this metric that guides patient treatment. Studies performed by NASA have assessed radiation dose dependence for specific symptoms. For example, it has been found that an exposure of 1.08 Gy is required to produce a 50% incidence of anorexia, an exposure of 1.58 Gy for nausea, and 2.4 Gy for vomiting. These studies have also quantified the time it takes for various symptoms to appear at specific doses. These metrics are particularly useful to an interplanetary crew, which will have no other methods of testing available. Once these metrics have been analyzed, the CMO will need to perform serial blood cell counts to determine white blood cell differentials, a metric that can be used to determine lymphocyte depletion.

In the first stage of the hematopoietic syndrome, neutrophils usually increase in response to the initial stress on the hematological system caused by the ionizing radiation. At the same time, lymphocytes decrease, and it is the ratio between the number of neutrophils and the number of lymphocytes that is useful to determine initial treatment. In addition to measuring the time it takes for an affected crewmember to start vomiting and analyzing lymphocyte depletion the CMO might, if the equipment is available, perform a cytogenetic dicentric assay [9]. This test, which is a criterion standard for biodosimetry, provides information about radiation-induced chromosomal aberrations. The only problem with this test is that it relies on using lymphocytes, and at exposure levels exceeding 5 Gy there will not be enough lymphocytes to perform the test. Another test that assesses genetic damage is the micronuclei formation assessment, a test that measures the formation of micronuclei caused by ionizing radiation. These extra micronuclei are caused by breaks in the chromosome, which provides an indication of the extent of the genetic damage [12].

The Reality of Treating ARS on a Spacecraft

Direct medical management of ARS will only be possible with access to laboratory diagnostics, since these are needed to determine the extent and magnitude of injury and illness. Serial blood work with differentials, serum amylase data, and cytogenetic biodosimetry are just a few of the diagnostics required for physicians to make an accurate diagnosis and prognosis and to manage injuries correctly. Even if these diagnostics were available it is unlikely symptomatic care—fluids, antiemetics, blood products, analgesics—could be provided to any crewmember except those who had suffered the lowest of low exposures.

References

1. Christensen DM, Iddins CJ, Parrillo SJ, Glassman ES, Goans RE. Management of ionizing radiation injuries and illnesses, part 4: radiation illness and injury: cutaneous radiation injury. J Am Osteopath Assoc. 2014;114:11.
2. Goans RE. Clinical care of the radiation-accident patient: patient presentation, assessment, and initial diagnosis. In: Ricks RC, Berger ME, O'Hara Jr FM, editors. The medical basis for radiation-accident preparedness: the clinical care of victims. Boca Raton: CRC Press; 2002.
3. Fliedner TM, Friesecke I, Beyrer K, editors. Medical management of radiation accidents: manual on the acute radiation syndrome. London: British Institute of Radiology; 2001. p. 13–38.
4. Dainiak N, Gent RN, Carr Z, et al. First global consensus for evidence-based management of the hematopoietic syndrome resulting from exposure to ionizing radiation. Disaster Med Public Health Prep. 2011;5(3):202–12.
5. Freifeld AG, Bow EJ, Sepkowitz KA, et al. Clinical practice guideline for the use of antimicrobial agents in neutropenic patients with cancer: 2010 update by the Infectious Diseases Society of America. Clin Infect Dis. 2011;52(4):e56-e93.
6. Baranov AE, Selidovkin GD, Butturini A, Gale RP. Hematopoietic recovery after 10-Gy acute total body radiation. Blood. 1994;83(2):596–9.
7. Zahorec R. Ratio of neutrophil to lymphocyte counts-rapid and simple parameter of systemic inflammation and stress in critically ill. Bratisl Lek Listy. 2001;102(1):5–14.
8. Zhang A, Azizova TV, Wald N, Day R. Changes of ratio of peripheral neutrophils and lymphocytes after radiation exposure may serve as a prognostic indicator of accident severity. In: Final Program, 49th Annual Meeting of the Health Physics Society. McLean, VA: Health Physics Society; 2004. Abstract P8:17.
9. Goans RE, Holloway EC, Berger ME, Ricks RC. Early dose assessment in criticality accidents. Health Phys. 2001;81(4):446–9.
10. Peter RU. Cutaneous radiation syndrome in multi-organ failure. BJR Suppl. 2005;27:180–4.
11. Goans RE. Cytogenetic biodosimetry and clinical hematology in the management of radiation injury. In: The medical basis for radiation-accident preparedness: medical management. Oak Ridge: Oak Ridge Associated Universities; 2013.
12. Stricklin D. Application of the premature chromosome condensation (PCC) assay in radiation dose assessment. In: The medical basis for radiation-accident preparedness: medical management. Oak Ridge: Oak Ridge Associated Universities; 2013. p. 379–89.

Chapter 9
Pharmacological Countermeasures

Space agencies conduct radiation research because astronauts are exposed to *chronic* doses of radiation. But during long duration missions beyond LEO there is a real danger that crews may be exposed to *acute* doses that may lead to acute radiation syndrome (ARS). To be prepared for such missions, space agencies must be prepared to anticipate radiation exposures and be able to deal with the consequences [1–4]. One way to do this is to implement a radiation medical countermeasures (Figs. 9.1 and 9.2) program that would cover products used following a radiological emergency.

Radiation Injury and Repair

One of the major targets for radiation is DNA, since it is a macromolecule, but a more abundant molecule is water, a key component of which is oxygen. The presence of oxygen is important because molecular oxygen is key to the formation of reactive free radicals, which means that in areas of high concentrations of oxygen, the effects of radiation are increased. Conversely, in areas of low oxygen concentration, tissues and cells are protected thanks to hypoxia. Of all the free radicals, the *hydroxyl* radical is one of the most damaging. One of the structures this free radical damages is DNA, a key structure for cell survival [5]. The hydroxyl radical, together with other oxidative radicals, are responsible for double-strand breaks (DSBs) and base lesions. Although the body goes to work repairing the damage, the process is not always successful since if two base lesions on opposite strands are too close, a double-strand break may occur [6]. This mechanism is known as the DNA damage response (DDR), and it is a process affected by the radiation dose, the type of radiation and the amount of tissue exposed to that radiation. The damaged DNA material is repaired thanks to the action of key proteins (MRE_{11}, RAD_{50} and NBS_1), which determine the damage before binding the broken DNA strands together in a process

E. Seedhouse, *Space Radiation and Astronaut Safety*, SpringerBriefs in Space Development, https://doi.org/10.1007/978-3-319-74615-9_9

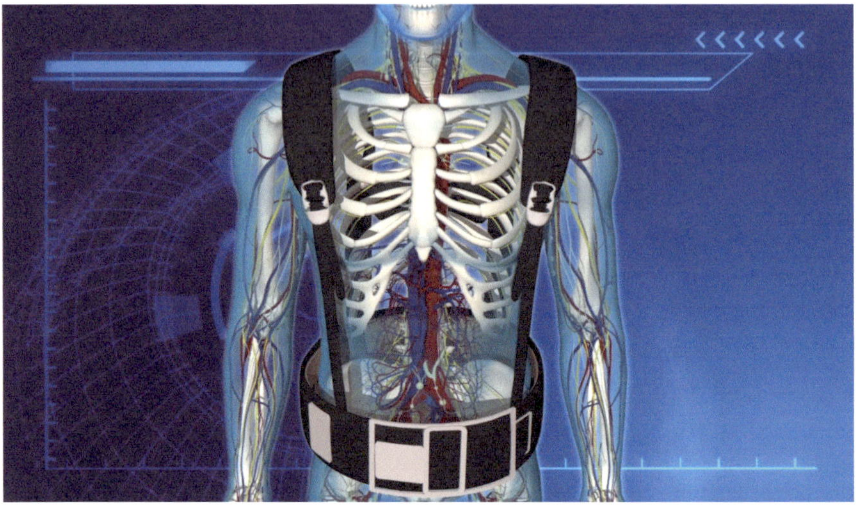

Fig. 9.1 StemRad—a possible way of protecting deep space astronauts from radiation (Image courtesy of NASA/StemRad)

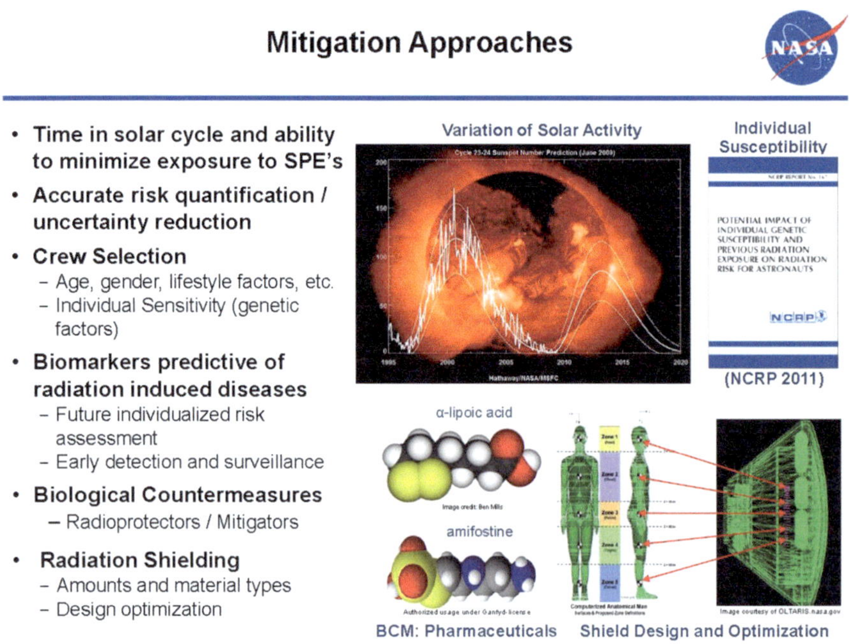

Fig. 9.2 NASA's radiation mitigation approaches. (Image courtesy of NASA)

that can proceed along two pathways. One of these pathways is homology-directed repair (HDR), which is a process that results in no errors. The other pathway is non-homologous end joining (NHEJ), which is a process that results in sequence deletions. If cells are not repaired successfully, genomic instability can be the result, which in turn can result in mutations and carcinogenesis. The extent of the mutations and the degree of risk of carcinogenesis will depend on the type of tissue, the radiation dose and the amount of tissue exposed to radiation [5]. How these mutations may occur is not fully understood, but it is probably a result of damage to progenitor cells, blood vessels and persistent oxidative stress. An extra risk is repeated exposure to radiation, which is definitely something deep space astronauts will be exposed to. In this case, repeated inflammation will be the result, which will lead to fibrosis and increased DNA damage, with possible outcomes being death, or at the very least permanent tissue damage [6].

Candidate Radioprotectors

Radioprotectors are compounds that can be considered as a pre-emptive medical countermeasure since they protect against radiation injury and the effects of ionizing radiation only when administered *before* any radiation exposure. This is different than a *mitigator*, which protects against radiation injury *after* exposure to radiation. Research that studies radioprotectors and mitigators usually investigate the effects of acute total-body irradiation (TBI) in rats. Although TBI affects several organ systems, death in the first 30 days, whether in rats or humans, is usually the result of two mechanisms (see Chap. 8):

1. Gastrointestinal Syndrome

   Death within 10–12 days after exposure of 8–20 Gy usually as a result of fluid and electrolyte imbalance and sepsis. In someone who is suffering from this syndrome, the fluid and electrolyte imbalance is caused by a depletion of intestinal stem cells, which are killed by the radiation in a process known as apoptosis [7].
2. Hematopoietic Syndrome

   Death within 30 days after exposure to 3–8 Gy, usually as a result of neutropenia and thrombocytopenia. In someone suffering from this syndrome, neutropenia and thrombocytopenia is caused by the depletion of radiosensitive hematopoietic progenitor cells for white blood cells [8].

To improve survival rates in astronauts who may be exposed to very high levels of radiation it is necessary to develop a radioprotector *and* mitigator that can protect against these syndromes. Ideally, such a compound should have a convenient mode of delivery and have low toxicity.

Amifostine

Unfortunately there are no radioprotectors or mitigators that have been approved for use in humans for preventing or treating the effects of acute radiation exposure. One agent used to reduce the toxicity of radiation therapy is Amifostine (Ethyol®), previously known as WR-2721 [9–11]. Developed by the U. S. Army Anti-Radiation Drug Development Program, Amifostine works thanks to a thiol compound that scavenges free radicals, thereby reducing the levels of oxidative radicals. Although it has been shown in studies in rats that Amifostine has some radioprotective effect, there are a number of limitations that include:

1. Narrow time window of administration. To have a radioprotective effect, Amifostine must be administered within 15–30 min before radiation exposure.
2. It has only been tested intravenously, although other routes may be possible.
3. Side effects. These include vomiting, nausea and hypotension [12, 13]. Not ideal for a crew of astronauts, although perhaps better than suffering the effects of acute radiation syndrome.

Superoxide Dismutase

Superoxide dismutase (SOD) has been subject to investigation for the transgene's ability to protect tissues against injury following radiation exposure. In mouse models, administration of this transgene did confer some protection against ulceration, and in mice that were fed a diet rich in antioxidants and administered SOD, lifespans were increased.

Genistein and Captopril

Genistein is a soy isoflavone that has been used as an anti-cancer agent [14–16]. It works by protecting bone marrow progenitor cells and reducing inflammation in tissues [17, 18]. Captopril, meanwhile, was originally developed to treat hypertension, but has since been investigated as a potential radiation countermeasure for the pulmonary and hematopoietic systems [19, 20]. How Captopril (Fig. 9.3) works is not completely understood, but research has shown that it blocks radiation-induced hematopoietic syndrome and reduces inflammation.

DBIBB

DBIBB was first highlighted following a study by Gábor Tigyi published in *Chemistry and Biology* in 2015 [21]. In Tigyi's study, DBIBB was shown to increase survival in mice exposed to radiation even after treatment had been administered

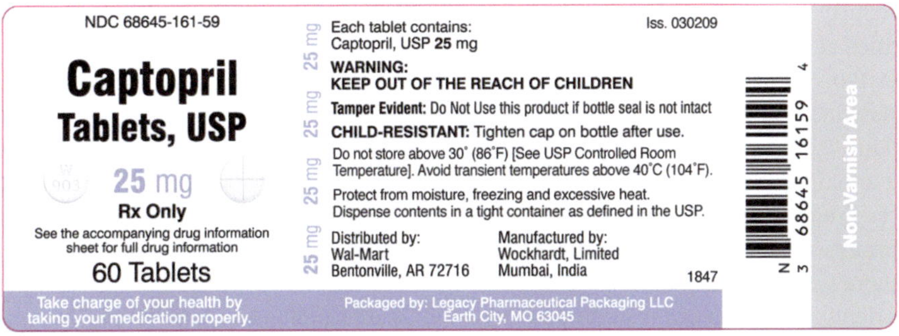

Fig. 9.3 Captopril. (Image courtesy of Legacy Pharmaceutical)

three days after exposure. In previous research, Dr. Tigyi and his colleagues had discovered that a molecule (lysophosphatidic acid, or LPA) generated during blood clotting, activates a receptor (called LPA_2) that protects against cell death caused by radiation. In this research scientists had also identified a compound similar to LPA that protected mice against radiation exposure. The problem with this compound was that it did not target the LPA2 receptor, and it was not potent enough to be used as a pharmacological countermeasure. So the researchers refined their study and engineered a more potent version of the LPA_2 receptor and dubbed it DBIBB. They then tested this compound in a mouse study and found that DBIBB increased the survival of radiation-exposed cells and protected DNA. In the study the group of mice that were not treated with DBIBB had a 20% survival rate, whereas the mice treated with DBIBB had a 93% survival rate. The next step was to test DBIBB on human hematopoietic progenitor cells. These cells were subject to radiation before being treated with DBIBB. In this study DBIBB significantly increased the survival of the cells. Although the study was one of a kind the results suggested that DBIBB could be the first radiomitigator.

Dietary Antioxidant Supplementation

Space radiation induces oxidative stress in cells, so it isn't surprising that scientists have suggested that astronauts might combat the effects of this stress by taking antioxidants. Oxidative stress occurs when there is a greater amount of pro-oxidants (radiation is a pro-oxidant) than antioxidants, and it is hypothesized that the use of antioxidants might counteract this imbalance [22, 23]. Scientists supporting this hypothesis argue that given the level of oxidative stress astronauts will be exposed to during exploration class missions, their intake of antioxidant vitamins will need to be significantly higher than recommended dietary allowances (RDAs).

One study [24] that tested the 'antioxidant as a radiation countermeasure' hypothesis, investigated an antioxidant supplement that contained a concoction of several antioxidant agents (ascorbic acid, co-enzyme Q10, α-lipoic acid, L-Selenomethionine, *N*-acetyl cysteine and vitamin E succinate) that were expected to reduce radiation-induced oxidative stress. The supplement was administered to mice at a weight basis equivalent to humans. One group of mice was then irradiated at the NASA Space Radiation Laboratory (NSRL) while another control group remained radiation free. Following their test both groups of mice were examined daily for 2 years for signs of toxicity such as ataxia, lack of grooming, weakness, anorexia, convulsions, twitching, tremors, bleeding, discharges, swelling, or labored respiration. Then, at the end of the 2-year period, the experimental and control groups were examined to measure any differences between the two groups. Since there were no statistically significant differences between the different diet groups, scientists were forced to conclude that antioxidant supplementation did not prevent the debilitating effects of radiation exposure. But there was some positive news because a more detailed analysis of the results revealed that antioxidant supplementation did prevent the more aggressive manifestations of radiation exposure such as malignant lymphomas and rare tumors.

Nicotinamide Mononucleotide

Nicotinamide mononucleotide (NMN) is a much hyped anti-aging drug that has been developed by scientists in Australia and the United States. NMN works by promoting DNA repair and could therefore help protect astronauts from radiation. NMD works by increasing levels of the oxidized form of nicotinamide adenine dinucleotide (NAD+), a chemical that is present in cells. NAD+ works by regulating protein interactions that help repair DNA, which is why NAD+ supplements have been very popular, although there has been little evidence that supports them having any anti-aging effect. NMN on the other hand, works so well that the scientists who performed the research are thinking of taking the drug themselves. In these studies, mice that were fed NMN supplements lived 20 percent longer than mice who were not fed the supplement. Of course, human trials have to be conducted, and assuming those trials are successful, the drug will need to be approved by the U. S. Food and Drug Administration.

How does NMN work? Well, as we age, our body's ability to repair itself becomes less and less efficient because the amount of NAD+ present in the cells declines and declines even more in those exposed to radiation. The theory is that if you can increase the amount of NAD+ in the cells you can enhance DNA repair. And the way you increase the amount of NAD+ in the cells is by adding a booster—NMN—that enhances the ability of cells to repair DNA. In some studies that have tested this theory the NMN not only increased the cells' ability to repair DNA but actually reversed existing genetic damage. And since it is predicted that about 5 percent of all the cells in an astronaut's body will die during a round-trip to Mars, NMN has caught the attention of scientists searching for ways to protect crewmembers during these missions.

Granulocyte Colony Stimulating Factor

As discussed in the previous chapter, one part of acute radiation syndrome is hematopoietic syndrome. This syndrome is characterized by a drop in the number of blood cells. This means that there is a reduction in the number of neutrophils, which is important, as these cells represent the first line of immune defense; as the numbers of neutrophils fall the risk of infection increases [25, 26]. Neutrophils are produced in the bone marrow from hematopoietic stem cells (HSCs), which give rise to multipotent progenitors (MPPs). The MPPs divide into mature blood cells via processes involving several regulators that are needed for maintaining homeostasis in the cells.

One of the key factors in the division, or differentiation, is granulocyte colony stimulating factor (G-CSF) [27]. Under normal conditions, most mature neutrophils stay in the bone marrow, and only 2% are released into the bloodstream as mature neutrophils. Once in the bloodstream, the differentiated neutrophils search for signs of infection, and if infection is detected, neutrophil chemoattractants are secreted, which trigger the production of G-CSF. (During infection, circulating neutrophils may increase by ten times the normal level.) But, as discussed earlier, when a person is exposed to high levels of ionizing radiation, neutrophil levels fall, resulting in neutropenia. Hypothetically, if it were possible to stimulate neutrophil levels by adding G-CSF, then perhaps neutropenia could be avoided and infection rates reduced. Such studies have been performed using pegfilgrastim (Neulasta®), which is a recombinant form of G-CSF, and filgrastim (Neupogen®) [27–29]. In one study, mice were irradiated to 2 Gy, and their neutrophil counts monitored for 30 days after exposure. Not surprisingly their neutrophil counts decreased significantly compared with the control group, which had not been irradiated. But when a second group of irradiated mice were administered filgrastim, the neutrophil counts returned to normal levels within two days. In another part of the study pegfilgrastim was administered to a control group of unirradiated mice, a procedure that boosted neutrophil counts by 15 times. Whether these effects would be observed in astronauts exposed to organ doses of 2 Gy is unknown, but if filgrastim and pegfilgrastim can show similar effects in humans the compounds may represent a mild countermeasure to radiation exposure.

Perspectives

Despite myriad research studies there are still no radioprotectors or mitigators available for long duration crews. There are some weak mitigators such as Vitamin E derivatives and there are compounds being tested on rodents that may be considered as weak radioprotectors, but none of these has been tested in human studies. So what are the characteristics of the ideal radioprotector/mitigator?

1. A weak radioprotector or mitigator is of little use for Mars-bound crews because of the ever present danger of solar particle events and the constant exposure to galactic cosmic radiation. A crew exposed to a high dose of radiation will need a radioprotector/mitigator capable of blocking radiation-induced mutagenesis and carcinogenesis [30, 31].
2. These agents will need to be effective for the first 24 h (or longer) following exposure.
3. They should have a convenient mode—intramuscular or subcutaneous—of administration.

References

1. Anno GH, Baum SJ, Withers HR, Young RW. Symptomatology of acute radiation effects in humans after exposure to doses of 0.5-30 Gy. Health Phys. 1989;56(6):821–38.
2. Waselenko JK, MacVittie TJ, Blakely WF, et al. Medical management of the acute radiation syndrome: recommendations of the Strategic National Stockpile Radiation Working Group. Ann Intern Med. 2004;140:1037–51.
3. Townsend LW. Implications of the space radiation environment for human exploration in deep space. Radiat Prot Dosim. 2005;115:44–50.
4. Cucinotta FA, Schimmerling W, Wilson JW, Peterson LE, Badhwar GD, Saganti PB, Dicello JF. Space radiation cancer risks and uncertainties for Mars missions. Radiat Res. 2001;156:682–8.
5. Slupphaug G, Kavli B, Krokan HE. The interacting pathways for prevention and repair of oxidative DNA damage. Mutat Res. 2003;531(1-2):231–51.
6. Lieber MR. The mechanism of double-strand DNA break repair by the nonhomologous DNA end-joining pathway. Annu Rev Biochem. 2010;79:181–211.
7. Booth C, Tudor G, Tudor J, Katz BP, MacVittie TJ. Acute gastrointestinal syndrome in high-dose irradiated mice. Health Phys. 2012;103(4):383–99.
8. Dainiak N. Hematologic consequences of exposure to ionizing radiation. Exp Hematol. 2002; 30(6):513–28.
9. Kouvaris JR, Kouloulias VE, Vlahos LJ. Amifostine: the first selective-target and broad-spectrum radioprotector. Oncologist. 2007;12(6):738–47.
10. van der Vijgh WJ, Peters GJ. Protection of normal tissues from the cytotoxic effects of chemotherapy and radiation by amifostine (Ethyol): preclinical aspects. Semin Oncol. 1994;21(5 Suppl 11):2–7.
11. Bourhis J, Blanchard P, Maillard E, Brizel DM, Movsas B, Buentzel J, et al. Effect of amifostine on survival among patients treated with radiotherapy: a meta-analysis of individual patient data. J Clin Oncol. 2011;29(18):2590–7.
12. Praetorius NP, Mandal TK. Alternate delivery route for amifostine as a radio-/chemo-protecting agent. J Pharm Pharmacol. 2008;60(7):809–15.
13. Seed TM, Inal CE, Singh VK. Radioprotection of hematopoietic progenitors by low dose amifostine prophylaxis. Int J Radiat Biol. 2014;90(7):594–604.
14. Verdrengh M, Jonsson IM, Holmdahl R, Tarkowski A. Genistein as an anti-inflammatory agent. Inflamm Res. 2003;52(8):341–6. 8
15. Landauer MR, Srinivasan V, Seed TM. Genistein treatment protects mice from ionizing radiation injury. J Appl Toxicol. 2003;23(6):379–85.
16. Para AE, Bezjak A, Yeung IW, Van Dyk J, Hill RP. Effects of genistein following fractionated lung irradiation in mice. Radiother Oncol. 2009;92(3):500–10.

17. Grace MB, Blakely WF, Landauer MR. Genistein-induced alterations of radiation-responsive gene expression. Radiat Meas. 2007;42:1152–7.
18. Ha CT, Li XH, Fu D, Xiao M, Landauer MR. Genistein nanoparticles protect mouse hematopoietic system and prevent proinflammatory factors after gamma irradiation. Radiat Res. 2013;180(3):316–25.
19. Davis TA, Landauer MR, Mog SR, Barshishat-Kupper M, Zins SR, Amare MF, et al. Timing of captopril administration determines radiation protection or radiation sensitization in a murine model of total body irradiation. Exp Hematol. 2010;38(4):270–81.
20. Moulder JE, Cohen EP, Fish BL, Hill P. Prophylaxis of bone marrow transplant nephropathy with captopril, an inhibitor of angiotensin-converting enzyme. Radiat Res. 1993;136(3):404–7.
21. Patil R, Szabó E, Fells JI, Balogh A, Lim KG, Fujiwara Y, Tigyi GJ. Combined mitigation of the gastrointestinal and hematopoietic acute radiation syndromes by a novel LPA_2 receptor-specific non-lipid agonist. Chem Biol. 2015;22(2):206–16.
22. Wan XS, Ware JH, Zhou Z, Donahue JJ, Kennedy AR. Protection against radiation induced oxidative stress in cultured human epithelial cells by treatment with antioxidant agents. Int J Radiat Oncol Biol Phys. 2006;64:1475–81.
23. Guan J, Wan XS, Zhou Z, Ware JH, Donahue JJ, Biaglow JE, Kennedy AR. The effects of dietary supplement agents on space radiation-induced oxidative stress in Sprague-Dawley rats. Radiat Res. 2004;162:572–9.
24. Guan J, Stewart J, Ware JH, Zhou Z, Donahue JJ, Kennedy AR. Effects of dietary supplements on the space radiation-induced reduction in total antioxidant status in CBA mice. Radiat Res. 2006;165:373–8.
25. Ware JH, Sanzari J, Avery S, et al. Effects of proton radiation dose, dose rate and dose fractionation on hematopoietic cells in mice. Radiat Res. 2010;174:325–30.
26. Romero-Weaver AL, Wan XS, Diffenderfer ES, et al. Kinetics of neutrophils in mice exposed to radiation and/or granulocyte colony-stimulating factor treatment. Radiat Res. 2013;180: 177–88.
27. Molineux G. The design and development of pegfilgrastim (PEG-rmetHuG-CSF, Neulasta). Curr Pharm Des. 2004;10(11):1235–44.
28. Scholz M, Ackermann M, Emmrich F, Loeffler M, Kamprad M. Effectiveness of cytopenia prophylaxis for different filgrastim and pegfilgrastim schedules in a chemotherapy mouse model. Biologics. 2009;3:27–37.
29. Fazzi R, Orciuolo E, Trombi L, Mattii L, Battola B, Riccioni R, et al. PEG-Filgrastim activity on granulocyte functions. Leuk Res. 2007;31(10):1453–5.
30. Kaur I, Simons ER, Castro VA, Mark Ott C, Pierson DL. Changes in neutrophil functions in astronauts. Brain Behav Immun. 2005;19(6):547–54.
31. Smith SM, Davis-Street JE, Rice BL, Nillen JL, Gillman PL, Block G. Nutritional status assessment in semiclosed environments: ground-based and space flight studies in humans. J Nutr. 2001;131:2053–61.

Chapter 10
Genetic Profiling and Manipulation

NASA plans to genetically modify astronauts so they can survive the journey to Mars.
– Dr. Douglas Terrier, Acting Chief Technologist at NASA

The prospect of humans one day visiting another planet has never seemed more possible than it has in recent years, but several more obstacles need to be overcome before this can become reality. Some people believe that incredible scientific discoveries could be waiting on the Red Planet, and NASA has recently ramped up its efforts in hopes of sending astronauts there in the 2030s.

Cosmic radiation is one of the biggest problems those who visit Mars will face, and it comes not only from the Sun but also stellar explosions that occur far beyond our Solar System. Particle radiation in space can collide with nuclei in human tissue, creating nuclear collisions that can create new particles and damage human cell DNA. Earth's magnetic field generally protects us from these particles, but the astronauts headed for Mars could be exposed to them for years without protection.

Now NASA is mulling over several ways to deal with this challenge, including the use of drugs that will alter the astronauts' DNA. They believe that this could help to repair any damage that is caused by the high-energy particles bombarding them when they visit the planet.

This approach was mentioned by Douglas Terrier, the agency's Acting Chief Technologist, ahead of an appearance at the London Codex Innovation Summit. Armored suits and shielded plating have also been suggested as possibilities but are believed to be impractical. In addition, NASA is looking into the possibility of using artificial intelligence programs to diagnose diseases in space and even carry out robotic surgery.

However, they are also considering altering the DNA of astronauts through epigenetic modifications, which will change the way the body reads genes without affecting the underlying DNA. By altering the chemicals controlling the volume of genes, their activity can be amplified or muted as needed should something go wrong.

E. Seedhouse, *Space Radiation and Astronaut Safety*, SpringerBriefs in Space Development, https://doi.org/10.1007/978-3-319-74615-9_10

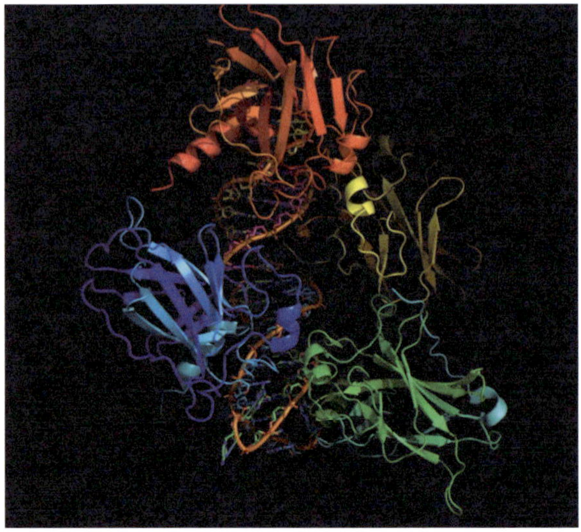

Fig. 10.1 Genetic manipulation may be one way of protecting future crews from radiation. (Image courtesy of NASA)

Dr. Terrier told *The Times*, "We're looking at a range of things. From drug therapies, and those seem to be quite promising, to more extreme things like epigenetic modification all the way to manipulation." (www.spacenews.com October 20, 2017. Isabelle Z.)

As we know, astronauts heading towards Mars will be facing myriad risks, chief among them being radiation. And one of the great uncertainties with such a mission is how much damage all that radiation will inflict. Perhaps shielding will help, perhaps not. Perhaps a faster propulsion system will help astronauts travel to Mars in 6 weeks rather than 6 months. The point is that scientists and mission planners need to look at *every* protective countermeasure on the table. And that may include genetically modifying humans. Risky? Probably. But so is a trip to Mars.

Some people think that radiation will keep NASA from sending people to Mars, but that's not the current situation. When we add the various mitigation techniques up, we are optimistic it will lead to a successful Mars mission with a healthy crew that will live a very long and productive life after they return to Earth, says Pat Troutman, NASA Human Exploration Strategic Analysis Lead (Fig. 10.1).

Troutman might be right, but the crew that does make that trip to Mars may not be baseline humans. The venerable space agency is doing its best to evaluate shielding materials, develop spacesuits to protect astronauts against radiation, investigate pharmaceutical countermeasures, integrate radiation sensors into spacecraft, and enhance radiation forecasting. But even with these efforts, all the positive and upbeat optimism in the world may not be enough to protect crews from radiation. So, what do we do? Well, there are a few options. One might be to select radiation resistant crews—providing we can identify a gene that indicates whether a person is resistant to radiation. Another option is to send clones. Yet another option is to genetically engineer astronauts to be radiation resistant (Figs. 10.2, 10.3, and 10.4).

Space Radiation Health Risks

Health Risk Areas	Status
Carcinogenesis Space radiation exposure may cause increased cancer morbidity or mortality risk in astronauts	➢ Cancer risk model developed for mission risk assessment ➢ Model is being refined through research at NASA Space Radiation Laboratory (NSRL) ➢ Health standard established
Acute Radiation Syndromes from SPEs Acute (in-flight) radiation syndromes, which may be clinically severe, may occur due to occupational radiation exposure	➢ Acute radiation health model has been developed and is mature ➢ Health standards established ➢ Risk area is controlled with operational & shielding mitigations
Degenerative Tissue Effects Radiation exposure may result in effects to cardiovascular system, as well as cataracts **Central Nervous System Risks (CNS)** Acute and late radiation damage to the central CNS may lead to changes in cognition or neurological disorders	➢ Non-cancer risks (**Cardiovascular and CNS**) are currently being defined ➢ Research is underway at NSRL and on ISS to address these areas ➢ Appropriate animal models needed to assess clinical significance

Fig. 10.2 Radiation health risks. (Image courtesy of NASA)

Post Mission Cancer Risk For A 900-day Mars Mission

Mars Mission Timing	Mission Shielding Configuration	Calculated REID, 95% C.I. (Age=45, Male-Female)	Amount Above 3% Standard
Solar Max	Good shielding like ISS (20 g/cm2) w/no exposure from SPEs	4% - 6%	1% - 3%
Solar Max	Good shielding like ISS (20 g/cm2) w/large SPE	5% - 7%	2% - 4%
Solar Min	Good shielding like ISS (20 g/cm2)	7% - 10%	4% - 7%

NASA Standards Limit The Additional Risk Of Cancer Death By Radiation Exposure, Not The Total Lifetime Risk Of Dying From Cancer

- Baseline lifetime risk of death from cancer (non-smokers)
  - 16% males, 12% females
- After Mars Mission (solar max), Astronauts lifetime risk of death from cancer ~20%

Fig. 10.3 Manned Mars mission radiation risks. (Image courtesy of NASA)

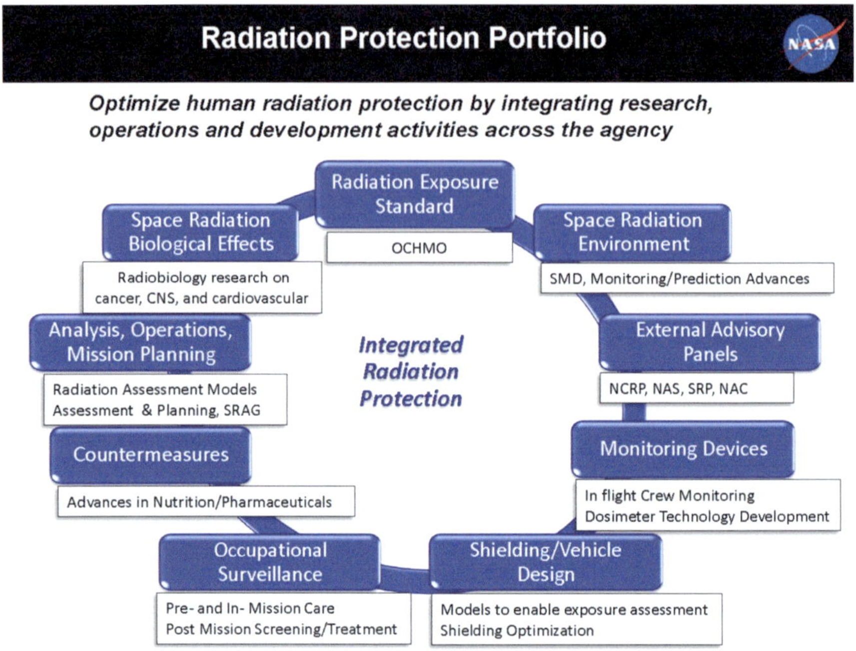

Fig. 10.4 NASA's radiation protection strategy. (Image courtesy of NASA)

The *Gattaca* Option

Gattaca[1] is a classic 1997 film starring Ethan Hawke, Uma Thurman and Jude Law. The film is set in a time when children are conceived via genetic manipulation to make sure they have only the best hereditary traits. The focus of the film is Ethan Hawke's character—Vincent Freeman—who was born outside of the eugenics program, which means he has to overcome genetic discrimination to realize his dream of becoming an astronaut. If the application of eugenics to create only the best astronauts sounds a little too close to Huxley's *Brave New World,* it is important to remember the novel foreshadowed many of the current technologies in reproductive science that we now take for granted. And while eugenics may conjure up dystopian visions, the term is derived from the Greek words '*eu*' which means good, and '*genos*,' which means offspring.

One of the reasons the science has a bad rap is because of the way the Nazis applied the ideas of eugenics. But recent advances in genetic engineering have opened the door for those who would like to alter genetic structure, a process known as *positive genetic engineering*. For astronauts deployed on a Mars mission, positive genetic engineering may prove a lifesaver. And that technology may not be too far

[1] The title is based on the letters that represent the four nucleobases of DNA, G (guanine), A (adenine), T (thymine) and C (cytosine).

away. For example, in 2002, researchers inserted insulin-like growth factor 1 (IGF-1) into the muscles of mice. This created the now familiar Schwarzenegger mice. In a similar study, scientists injected mice with the fat-burning protein PPAR-δ, a procedure that allowed mice to become endurance athletes. So why not manipulate the genes of astronauts so they can be more radiation resistant? Would it really be so bad if scientists tweaked the genes of future spacefarers so they wouldn't have to worry about radiation?

Extremophiles

If scientists are going to upgrade astronauts to be radiation resistant they will need to find radiation resistant genes. Fortunately, there are several organisms that have evolved mechanisms to ensure genetic integrity even in the face of extreme radiation. This class of organisms are known as *extremophiles*.

Deinococcus radiodurans

One such extremophile is *Deinococcus radiodurans*, a bacterium that can survive extreme levels of ionizing radiation [1–3]. Before we take a closer look, it is worth reviewing the effects of ionizing radiation on genetic material. The first type of damage is direct damage. This occurs when DNA absorbs ionizing radiation, which in turn leads to single and double strand breaks (DSBs). Indirect damage, on the other hand, occurs when reactive oxygen species (ROS)[2] interact with DNA. For example, hydroxyl radicals, a type of ROS, can be formed when ionizing radiation is absorbed by water [4]. Under these circumstances, hydroxyl radicals can cause single strand breaks (SSBs) and when two of these SSBs are close together there is the risk of a DSB forming. And it is the DSB breaks that are so damaging because they put a stop to DNA replication and are very difficult to repair.

Having said that, the body can repair DSBs, and it can do this by using recombinational DNA repair and end joining. This repair capability varies, with some genes more able to repair than others, and scientists think that one of the keys to this capability is linked to cell survival after the exposure to ionizing radiation. To get a better idea of how this process works, scientists have studied radiation-resistant extremophiles such as *Deinococus radiodurans,* and these studies have found certain adaptations to genetic repair mechanisms [5, 6]. Unfortunately, due to the complexity of bacterial metabolism it has been difficult for scientists to isolate these mechanisms,

[2] ROS includes chemical species such as superoxides and hydroxyl radicals that are formed as a by-product of normal oxygen metabolism. The ROS play important roles in cell homeostasis, but during environmental stress, ROS levels are increased. This increase leads to cell damage, and it is this damage that is known as oxidative stress.

but research continues in an attempt to understand how these bacteria can survive lethal radiation. For example, *Deinococus radiodurans* can survive doses greater than 5 kGy, which is more than 1000 times the lethal dose for humans. How can the bacterium do this? Well, scientists think there are three mechanisms that can explain this level of radiation resistance. One is a specialized nucleoid structure that may accelerate genetic repair processes. A second is a greater capacity for repairing protein damage that would better protect DNA, and a third is a specialized pathway for DNA repair.

The details of this research are beyond the scope of this book. Suffice to say that extremophiles such as *Deinococus radiodurans* have a toolbox of genetic repair pathways, and these have evolved as an adaptive response to a harsh environment. So how did these bacteria *acquire* this radiation resistance? Well, in an attempt to answer that question, scientists bombarded radiation-sensitive bacteria with increasingly high levels of radiation to see if and how populations adapted. They started by subjecting a population of radiation-sensitive bacteria to radiation levels (3000 Gy) high enough to kill 99% of the population [7]. This procedure was then repeated in four separately evolved populations. Scientists found that the bacteria evolved to resist the radiation by employing a pattern of genetic innovation that is comprised of three functional categories: DNA repair and replication, oxidative damage suppression and cell wall biogenesis. A detailed explanation of the mechanisms involved in these adaptive processes is beyond the scope of this book, but it is worth briefly discussing two of them.

One key process scientists discovered was the malleability of bacterial DNA repair systems. Rather than just facilitate more efficient repair in one pathway, the bacteria adapted by facilitating multiple radiation resistance pathways [8, 9]. Another key to the evolution of these bacteria was the rapid speed of adaptation. This, combined with the range of repair and recovery mechanisms conjured up by these bacteria, added up to the ability to acquire extreme resistance to radiation. In short, multiple pathways combined with multiple adaptations explain how bacteria such as *Deinococus radiodurans* can adapt its genes to be less vulnerable to radiation

Omics

Omics describes a personalized medicine paradigm. Today's astronauts undergo the most advanced medical selection imaginable, but they are not required to undergo phenotyping, metabolic network assessment or tests to see how radiation resistant they are. But that may soon change. That's because personalized medicine—Omics—may be one of the keys to protecting astronauts from radiation. Applied to manned spaceflight Omics would apply information about an astronaut's genes in the context of that astronaut's diet, lifestyle and environment with the goal of optimizing safety and performance commensurate with mission demands [10]. Generally speaking, the Omics paradigm takes advantage of advances in genomics, transcriptomics, proteomics, metabolomics, medical treatment and bioinformatics to better characterize individual profiles. Of these areas, it is genomics and medical treatment that is of interest to those charged with keeping astronauts safe from radiation. In short, Omics is an applied systems biology approach to space biomedicine. The good news is that Omics is in use today. For example, doctors use diagnostic tests (Onco*type* DX® and MammaPrint®) together with genetic information to decide how best to treat breast cancer. So how might Omics be applied to selecting/screening astronauts? Well, to explain the scope of Omics as it might be applied to astronaut selection is beyond the scope of this publication, so what follows is a brief insight.

One Carbon Metabolism Variation

One carbon metabolism is linked to genetic repair mechanisms, and therefore variations in one carbon metabolism directly affect how resistant an astronaut is to radiation effects. One marker for this within the one carbon metabolism mechanism is uracil [11, 12]. If levels of this enzyme are too high the result is a higher likelihood of single-strand breaks. So Omics may be used to identify one carbon deficits and by so doing perhaps identify astronauts who have a higher or lower susceptibility to chromosomal damage. Incidentally such an approach may also help identify astronauts more or less resistant to ocular disturbances and bone health [13–15] since variants in one carbon metabolism are linked to eye health and skeletal strength.

HFE and Oxidative Stress

Another Omics application may be to measure iron parameters [16, 17] in the HFE (the hemochromatosis gene) group, since this may be related to oxidative stress, and that in turn is related to genetic repair mechanisms. As previously discussed we know genetic material is very sensitive to oxidative stress, and this stress can be triggered by exposure to iron. Studies that have investigated this link have done so in those suffering from hemochromatosis, which is a hereditary condition. Those who have hemochromatosis live with an iron overload that contributes to increased oxidative stress, so it would make sense to measure levels of HFE in astronauts to see if it might serve as a biomarker for oxidative stress.

Micronutrients

Omics research has revealed that certain small molecules can exert a significant effect on the human response to spaceflight. These small molecules are given the term *essential inputs,* and this class of molecules can, depending on the astronaut's genotype, exert a significant effect on how that astronaut is affected by the environment. For example, niacin is involved in hundreds of metabolic reactions, including those affecting genomic stability, genetic repair and chromosomal damage. Likewise zinc plays an important role in protecting genetic material from damage and is also involved in DNA transcription, regulation and repair. In short, if an astronaut's niacin and zinc levels are too low they will be more susceptible to DNA damage.

Selenium is another essential input that may serve a protective purpose in preventing genetic damage by enhancing the activity of repair enzymes [18, 19]. Magnesium also plays a role in maintaining the health of DNA. Specifically, magnesium works as a cofactor for enzyme systems that assist in repairing DNA, mismatch repair and nucleotide excision repair [20]. Mismatch repair is a mechanism that recognizes bases that have not been repaired properly. This mechanism also recognizes recombination between non-identical DNA sequences. When this mismatch repair mechanism is working normally it helps to significantly lower the mutation frequency in the genome, but if an astronaut's magnesium levels are too low the mismatch repair mechanism is impaired.

Aggregation of Essential Inputs

One can easily see that the aggregation of one or more variants in essential inputs may have a pronounced effect on DNA health. For example, if an astronaut had an altered one carbon metabolism then he or she might be at a higher risk of unstable DNA or strand breaks. And if that same astronaut also had a magnesium deficiency then he or she would have an impaired efficiency to repair that damage. This scenario is not too farfetched because altered iron metabolism has been shown to exist in the astronaut population. By applying Omics to astronauts with the purpose of identifying off-target influences and finding biomarkers that may limit the ability of astronauts to participate in missions, it will be possible to develop personalized countermeasures and also to profile the ideal genotype. Such an approach will result in the development of an individualized countermeasure program that is personalized, preventive and predictive [21].

References

1. Blasius M, Hubscher U, Sommer S. Deinococcus radiodurans: what belongs to the survival kit? Crit Rev Biochem Mol Biol. 2008;43:221–38.
2. Cox MM, Battista JR. Deinococcus radiodurans - the consummate survivor. Nat Rev Microbiol. 2005;3:882–92.
3. Daly MJ. Opinion: a new perspective on radiation resistance based on Deinococcus radiodurans. Nat Rev Microbiol. 2009;7:237–45.
4. Slade D, Radman M. Oxidative stress resistance in Deinococcus radiodurans. Microbiol Mol Biol Rev. 2011;75:133–91.

5. Krisko A, Radman M. Protein damage and death by radiation in Escherichia coli and Deinococcus radiodurans. Proc Natl Acad Sci U S A. 2010;107:14373–7.
6. Levin-Zaidman S, Englander J, Shimoni E, Sharma AK, Minton KW, Minsky A. Ringlike structure of the Deinococcus radiodurans genome: a key to radio-resistance? Science. 2003; 299:254–6.
7. Zahradka K, Slade D, Bailone A, Sommer S, Averbeck D, Petranovic M, Lindner AB, Radman M. Reassembly of shattered chromosomes in Deinococcus radiodurans. Nature. 2006;443:569–73.
8. Slade D, Lindner AB, Paul Gradman M. Recombination and replication in DNA repair of heavily irradiated Deinococcus radiodurans. Cell. 2009;136:1044–55.
9. Daly MJ, Gaidamakova EK, Matrosova VY, Vasilenko A, Zhai M, Venkateswaran A, Hess M, Omelchenko MV, Kostandarithes HM, Makarova KS, Wackett LP, Fredrickson JK, Ghosal D. Accumulation of Mn(II) in, Deinococcus radiodurans facilitates gamma-radiation resistance. Science. 2004;306:1025–8.
10. Tanaka H. Omics-based medicine and systems pathology: a new perspective for personalized and predictive medicine. Methods Inf Med. 2010;49:173–85.
11. Fredriksen A, Meyer K, Ueland PM, et al. Large-scale population-based metabolic phenotyping of thirteen genetic polymorphisms related to one-carbon metabolism. Hum Mutat. 2007;28:856–65.
12. Vijaya Lakshmi SV, Naushad SM, Seshagiri Rao D, Kutala VK. Oxidative stress is associated with genetic polymorphisms in one-carbon metabolism in coronary artery disease. Cell Biochem Biophys. 2011;67(2):353–61. https://doi.org/10.1007/s12013-011-9322-1.
13. Smith SM, Heer M. Calcium and bone metabolism during space flight. Nutrition. 2002;18: 849–52.
14. Smith SM, Heer MA, Shackelford LC, Sibonga JD, Ploutz-Snyder L, et al. Benefits for bone from resistance exercise and nutrition in long-duration spaceflight: evidence from biochemistry and densitometry. J Bone Miner Res. 2012;27:1896–906.
15. Zwart SR, Morgan JLL, Smith SM. Iron status and its relations with oxidative damage and bone loss during long-duration space flight on the International Space Station. Am J Clin Nutr. 2013;98(1):217–23.
16. Nandar W, Connor JR. HFE gene variants affect iron in the brain. J Nutr. 2011;141:729S–39S.
17. Jahanshad N, Kohannim O, Hibar DP, et al. Brain structure in healthy adults is related to serum transferrin and the H63D polymorphism in the HFE gene. Proc Natl Acad Sci U S A. 2012;109:E851–9.
18. Bera S, Rosa VD, Rachidi W, Diamond AM. Does a role for selenium in DNA damage repair explain apparent controversies in its use in chemoprevention? Mutagenesis. 2013;28(2):127–34.
19. Bianco AC, Larsen PR. Selenium, deiodinases and endocrine function. In: Hatfield DL, Berry MI, Gladyshev VN, editors. Selenium: its molecular biology and role in human health. 2. New York: Springer; 2006. p. 207–19.
20. Rowe WJ. Long space missions, gene therapy, and the vital role of magnesium: A three-pronged plan for the next 50 years. Int J Nephrol Renov Dis. 2010;3:123–7.
21. Personalized Medicine Coalition. The case for personalized medicine. 2010. Washington, DC. www.personalizedmedicine.org.

Correction to: Radiation Dosimetry and Detection

Correction to: Chapter 5 in: E. Seedhouse, *Space Radiation and Astronaut Safety*, SpringerBriefs in Space Development, https://doi.org/10.1007/978-3-319-74615-9_5

The original version of this chapter has been revised because it was inadvertently published with the below errors. The initially published version has now been corrected.

Chapter 5

On pages 54 and 55, Figures 5.5 and 5.6: Copyright indicated as NASA has been changed to ESA.

On page 54. The acronym 'Escaped' has been corrected to 'EuCPAD'.

The updated original version of this chapter can be found at https://doi.org/10.1007/978-3-319-74615-9_5

E. Seedhouse, *Space Radiation and Astronaut Safety*, SpringerBriefs in Space Development, https://doi.org/10.1007/978-3-319-74615-9_11

Appendix A

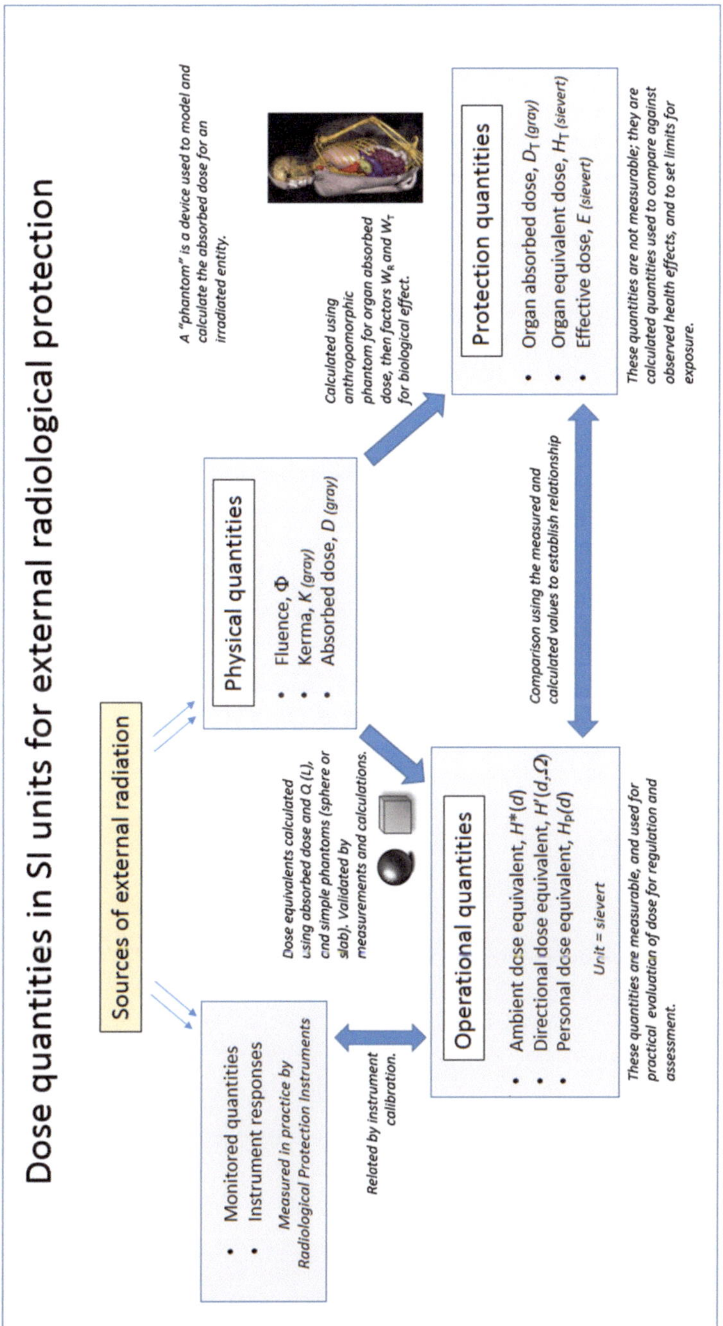

E. Seedhouse, *Space Radiation and Astronaut Safety*, SpringerBriefs in Space Development, https://doi.org/10.1007/978-3-319-74615-9

Appendix B

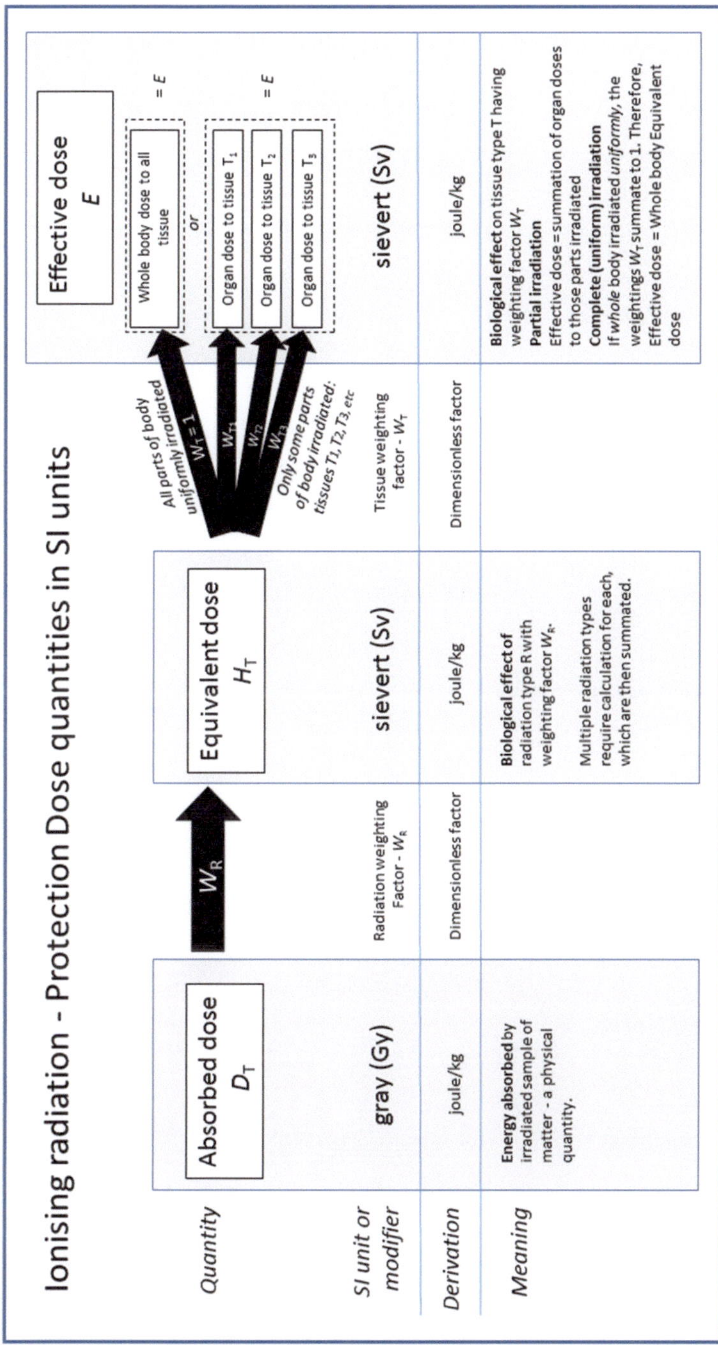

E. Seedhouse, *Space Radiation and Astronaut Safety*, SpringerBriefs in Space Development, https://doi.org/10.1007/978-3-319-74615-9

Appendix C

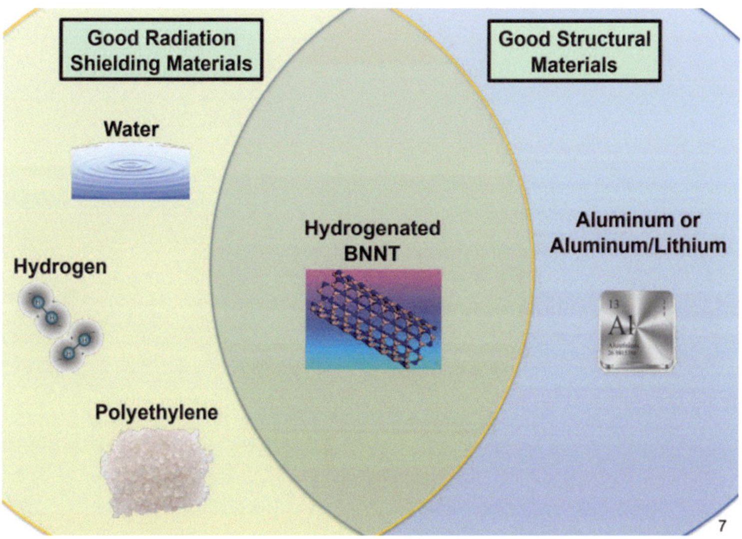

E. Seedhouse, *Space Radiation and Astronaut Safety*, SpringerBriefs in Space Development, https://doi.org/10.1007/978-3-319-74615-9

Index

E. Seedhouse, *Space Radiation and Astronaut Safety*, SpringerBriefs in Space Development, https://doi.org/10.1007/978-3-319-74615-9

MIX
Papier aus verantwortungsvollen Quellen
Paper from responsible sources
FSC® C105338

If you have any concerns about our products,
you can contact us on
ProductSafety@springernature.com

In case Publisher is established outside the EU,
the EU authorized representative is:
Springer Nature Customer Service Center GmbH
Europaplatz 3, 69115 Heidelberg, Germany

Printed by Libri Plureos GmbH
in Hamburg, Germany